W9-AXW-056

McGraw-Hill's Math

GRADE 1

New York Chicago San Francisco Athens London Madrid
Mexico City Milan New Delhi Singapore Sydney Toronto

Copyright © 2012 by McGraw-Hill Education. All rights reserved. Printed in the United States of America. Except as permitted under the United States Copyright Act of 1976, no part of this publication may be reproduced or distributed in any form or by any means, or stored in a database of retrieval system, without the prior written permission of the publisher.

3 4 5 6 7 8 9 10 11 12 13 14 15 LWI 21 20 19 18 17 16

ISBN 978-0-07-177556-4
MHID 0-07-177556-0

e-ISBN 978-0-07-177557-1
e-MHID 0-07-177557-9

Cataloging-in-Publication data for this title are on file at the Library of Congress.

Library of Congress Control Number: 2011914996

Printed and bound by LSC Communications.

Editorial Services: Pencil Cup Press
Production Services: Jouve
Illustrator: Eileen Hine
Designer: Ella Hanna

McGraw-Hill Education books are available at special quantity discounts to use as premiums and sales promotions or for use in corporate training programs. To contact a representative, please visit the Contact Us pages at www.mhprofessional.com.

This book is printed on acid-free paper.

Table of Contents

Table of Contents

Chapter 7

Chapter 8

Chapter 9

Chapter 10

Chapter 11

Chapter 12

Welcome to McGraw-Hill's Math!

This is your math book.
Its lessons tell math ideas.
The lessons give practice in using skills too.

Open your book. Look at the Table of Contents.
It shows the math ideas in each lesson.

Then look at the 10-Week Summer Study Plan.
It shows one way to plan your time.
But you may want to work faster or more slowly.

Each group of lessons ends with a Chapter Test.
Take the test.
It will show you how well you can do the math.

There are 2 Reviews in this book.
Complete these pages.
They will show you how much you have learned.

10-Week Summer Study Plan

Many children will use this book as a summer study program.
Use this 10-week study plan to help plan the time.
Put a ✔ in the box when the child finishes the day's work.

	Day	Lesson Pages	Test Pages
Week 1	Monday	8, 9	
	Tuesday	10, 11	
	Wednesday	12, 13	
	Thursday		14–15
	Friday	16	
Week 2	Monday	17, 18	
	Tuesday	19, 20	
	Wednesday	21	22–23
	Thursday	24, 25	
	Friday	26, 27	
Week 3	Monday	28, 29	
	Tuesday		30–31
	Wednesday	32, 33	
	Thursday	34, 35	
	Friday		36–37
Week 4	Monday	38, 39	
	Tuesday	40	
	Wednesday	41	
	Thursday	42	43–44
	Friday	45	
Week 5	Monday	46, 47	
	Tuesday	48	
	Wednesday	49	
	Thursday		50–51
	Friday		REVIEW 52–55

	Day	Lesson Pages	Test Pages
Week 6	Monday	56, 57	
	Tuesday	58, 59	
	Wednesday	60, 61	
	Thursday	62, 63	
	Friday	64	65–66
Week 7	Monday	67, 68	
	Tuesday	69, 70	71–72
	Wednesday	73	
	Thursday	74, 75	
	Friday	76, 77	
Week 8	Monday	78, 79	
	Tuesday		80–81
	Wednesday	82, 83	
	Thursday	84, 85	
	Friday	86, 87	
Week 9	Monday	88, 89	90–91
	Tuesday	92, 93, 94	
	Wednesday	95, 96	
	Thursday	97, 98	
	Friday	99, 100	
Week 10	Monday	101, 102	
	Tuesday	103, 104	105–106
	Wednesday	107, 108, 109	
	Thursday	110	111–112
	Friday		REVIEW 113–118

Name ____________________

Counting and Writing from 0 to 5

You can count to find out how many.

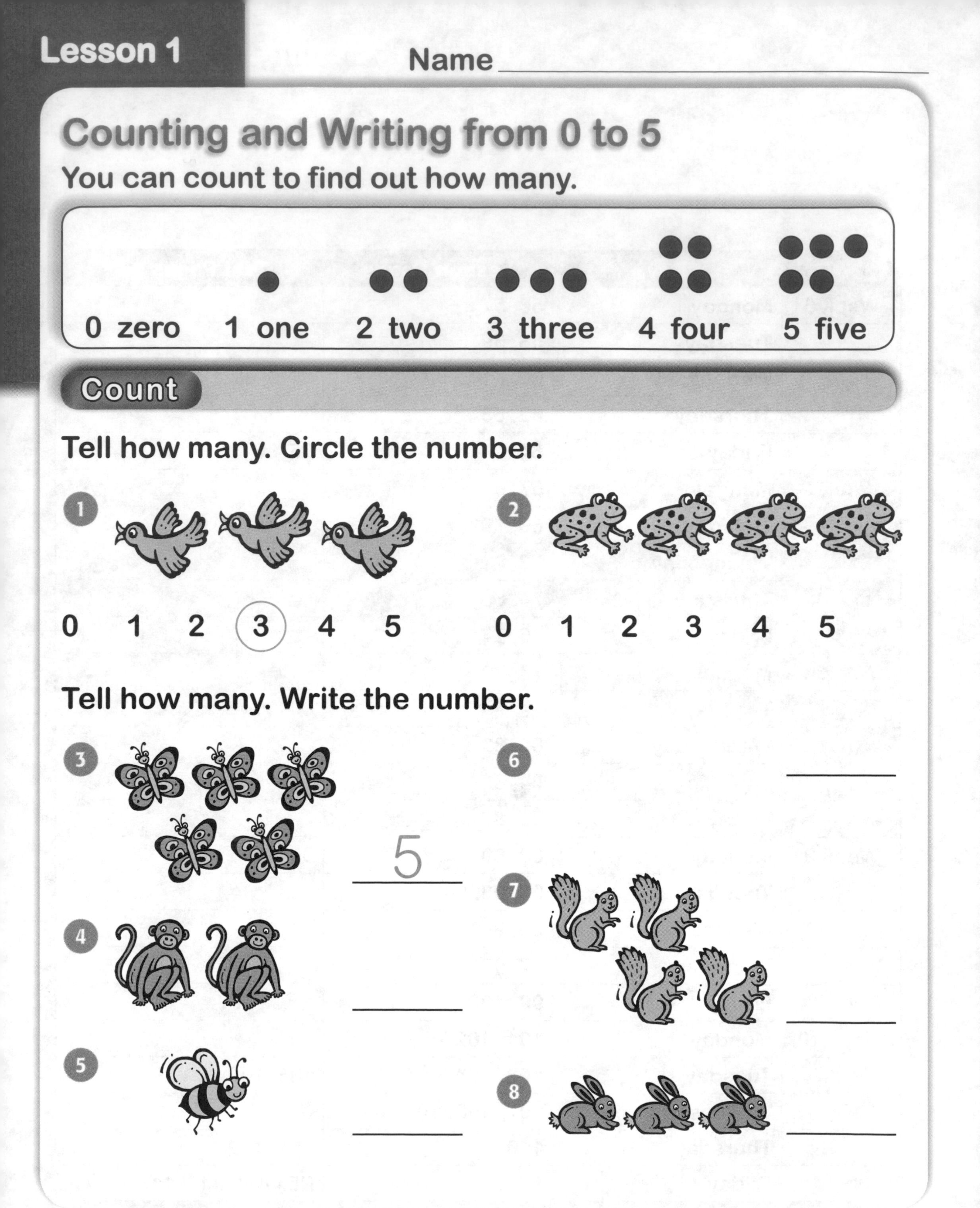

Tell how many. Circle the number.

Tell how many. Write the number.

Name ____________________

Counting and Writing from 6 to 10

You can count to find out how many.

6 six 7 seven 8 eight 9 nine 10 ten

Count

Tell how many. Circle the number.

1. 6 (7) 8 9 10

2. 6 7 8 9 10

Tell how many. Write the number.

3. 8

4. ______

5. ______

6. ______

7. ______

8. ______

Name ____________________

Counting and Writing from 11 to 15

You can write numbers to tell how many.

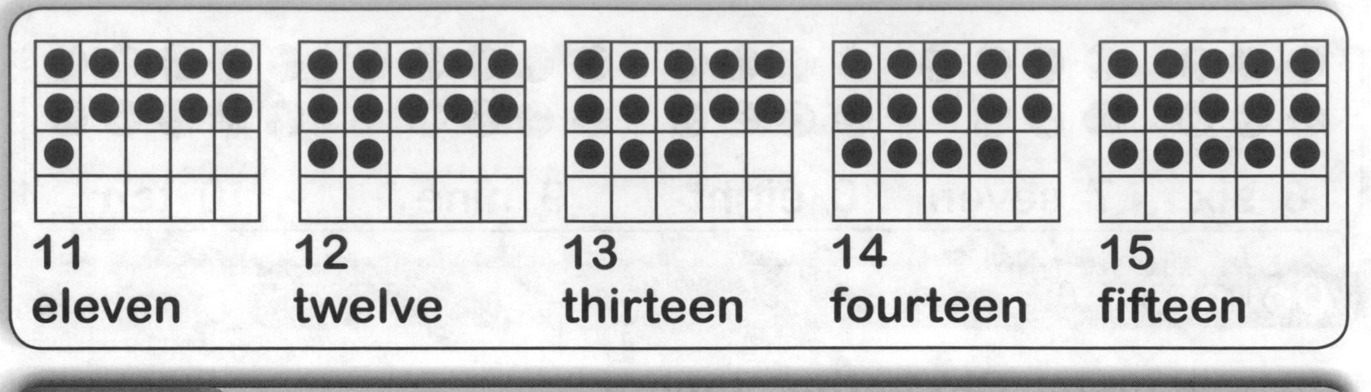

Count

Tell how many. Circle the number.

1. 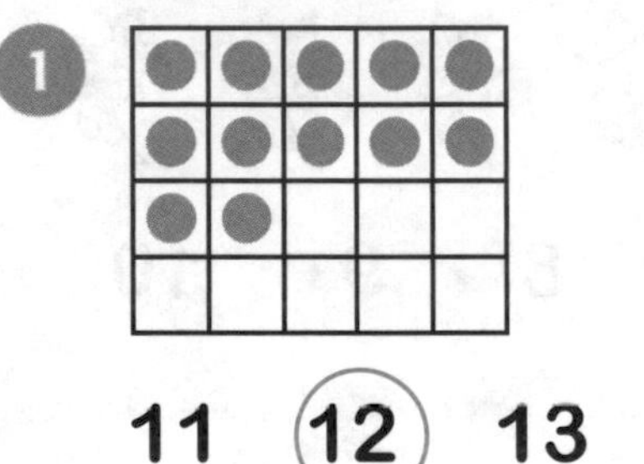

11 (12) 13 14 15

2. 11 12 13 14 15

Tell how many. Write the number.

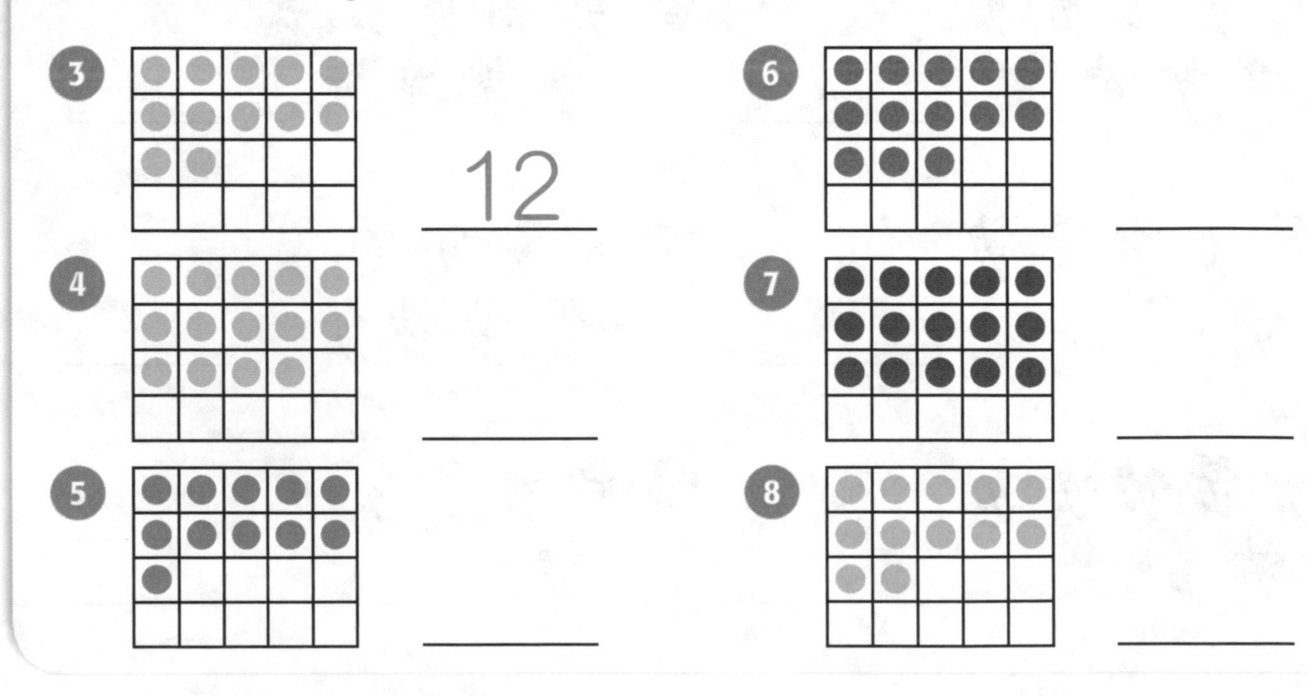

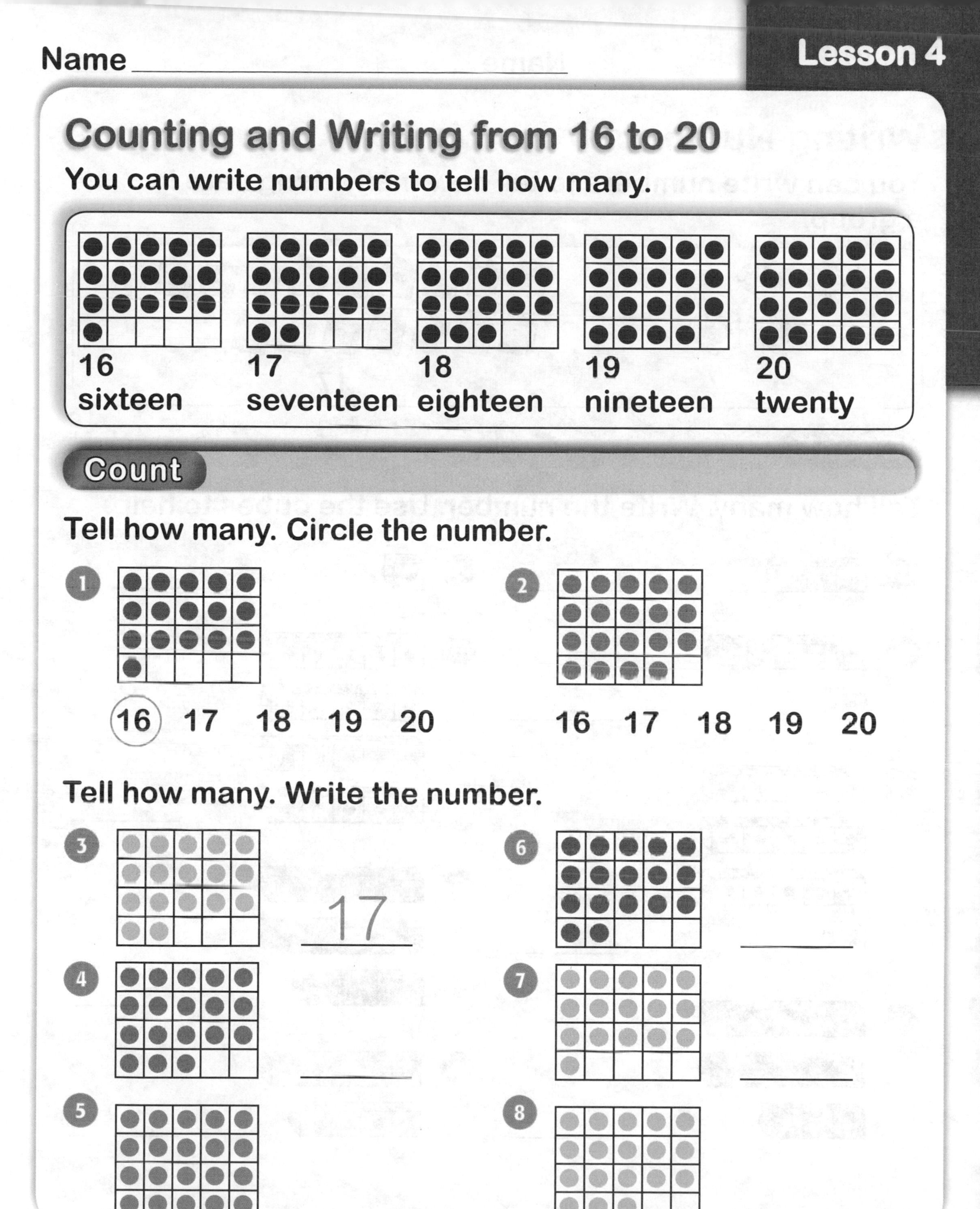

Name ______________________

Lesson 4

Counting and Writing from 16 to 20

You can write numbers to tell how many.

16	17	18	19	20
sixteen	seventeen	eighteen	nineteen	twenty

Count

Tell how many. Circle the number.

1. 16 17 18 19 20

2. 16 17 18 19 20

Tell how many. Write the number.

3. 17

4. ______

5. ______

6. ______

7. ______

8. ______

Name ______________________

Writing Numbers from 0 to 20

You can write numbers to tell how many things are in a group.

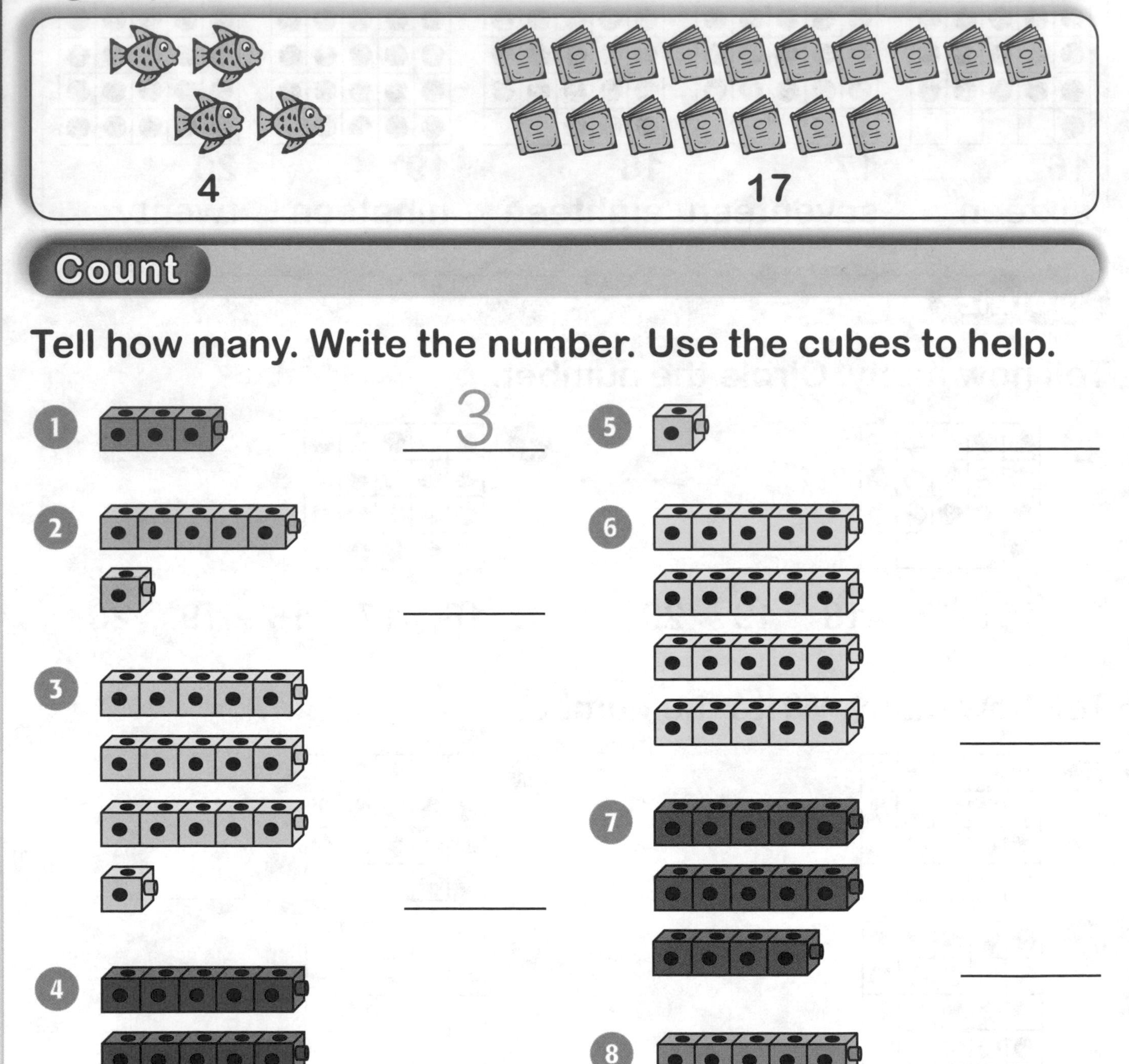

Count

Tell how many. Write the number. Use the cubes to help.

1. 3
2. ______
3. ______
4. ______
5. ______
6. ______
7. ______
8. ______

Name ______________________

Showing Numbers of Things

You can write numbers to tell how many things are in a group.

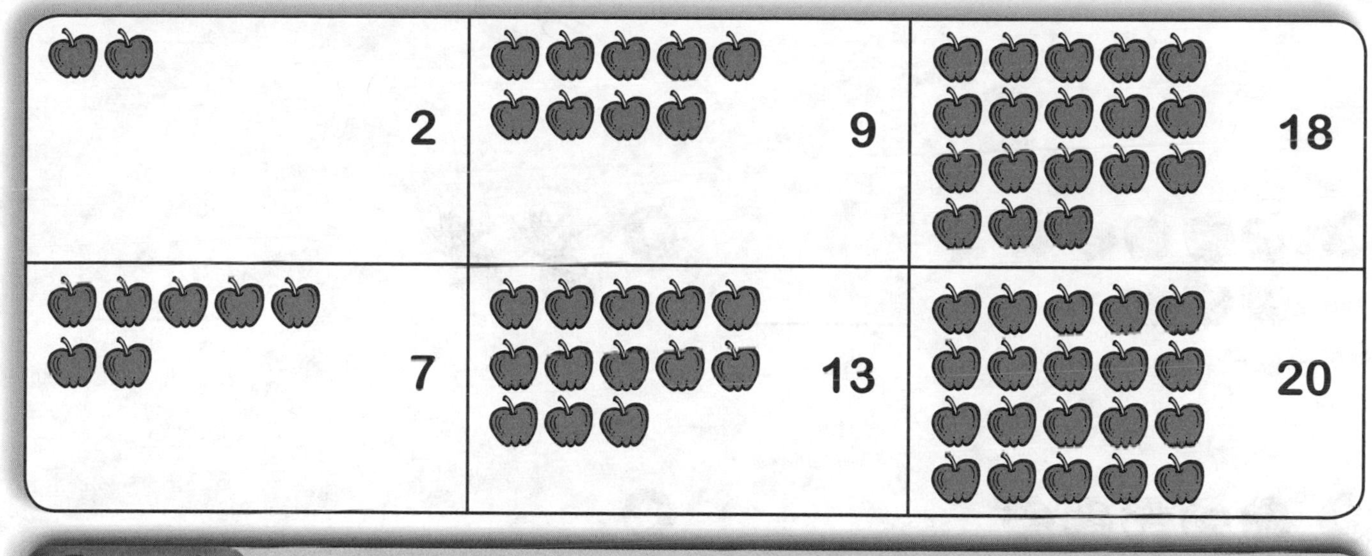

Count

Tell how many. Write the number.

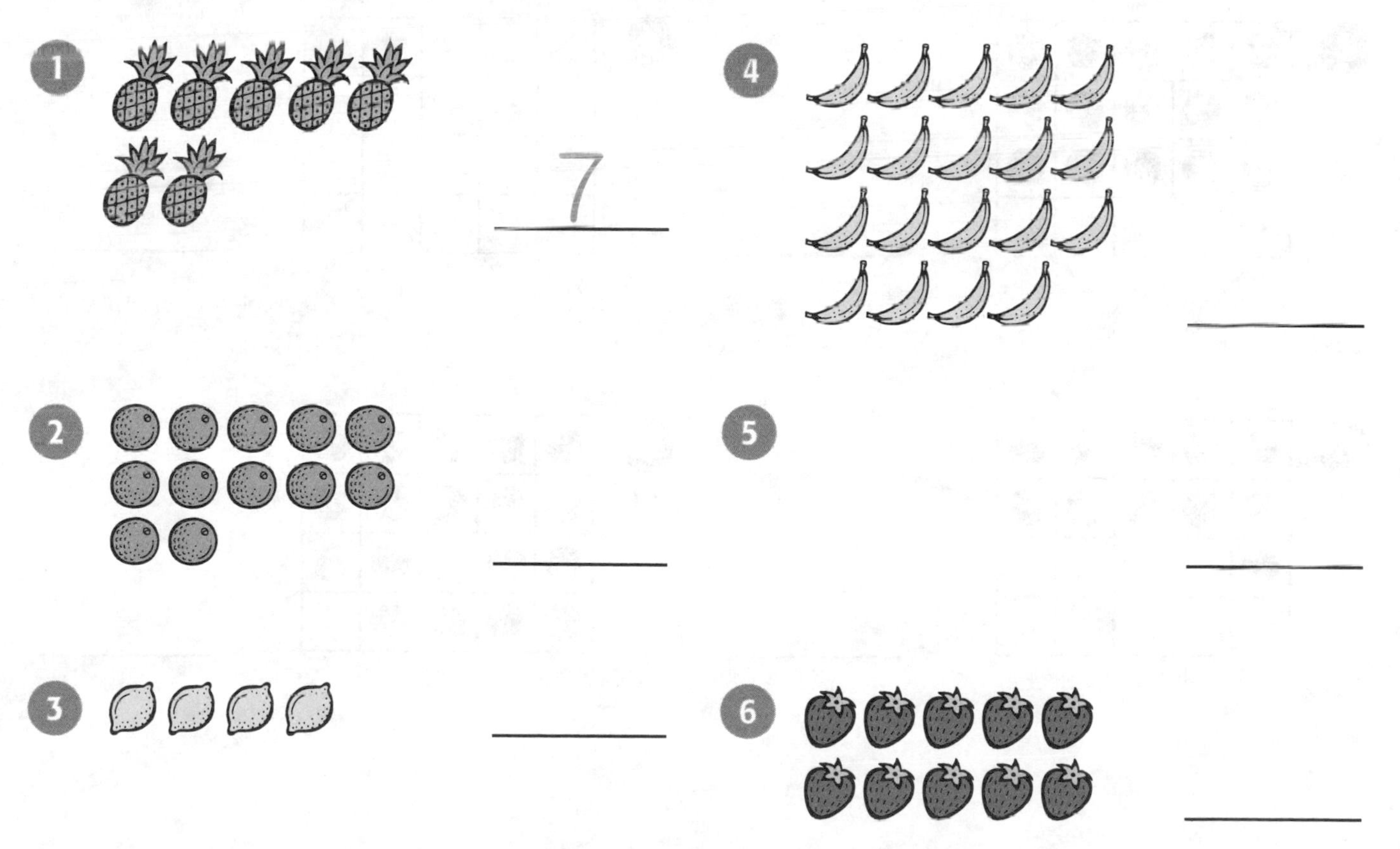

Name ____________________

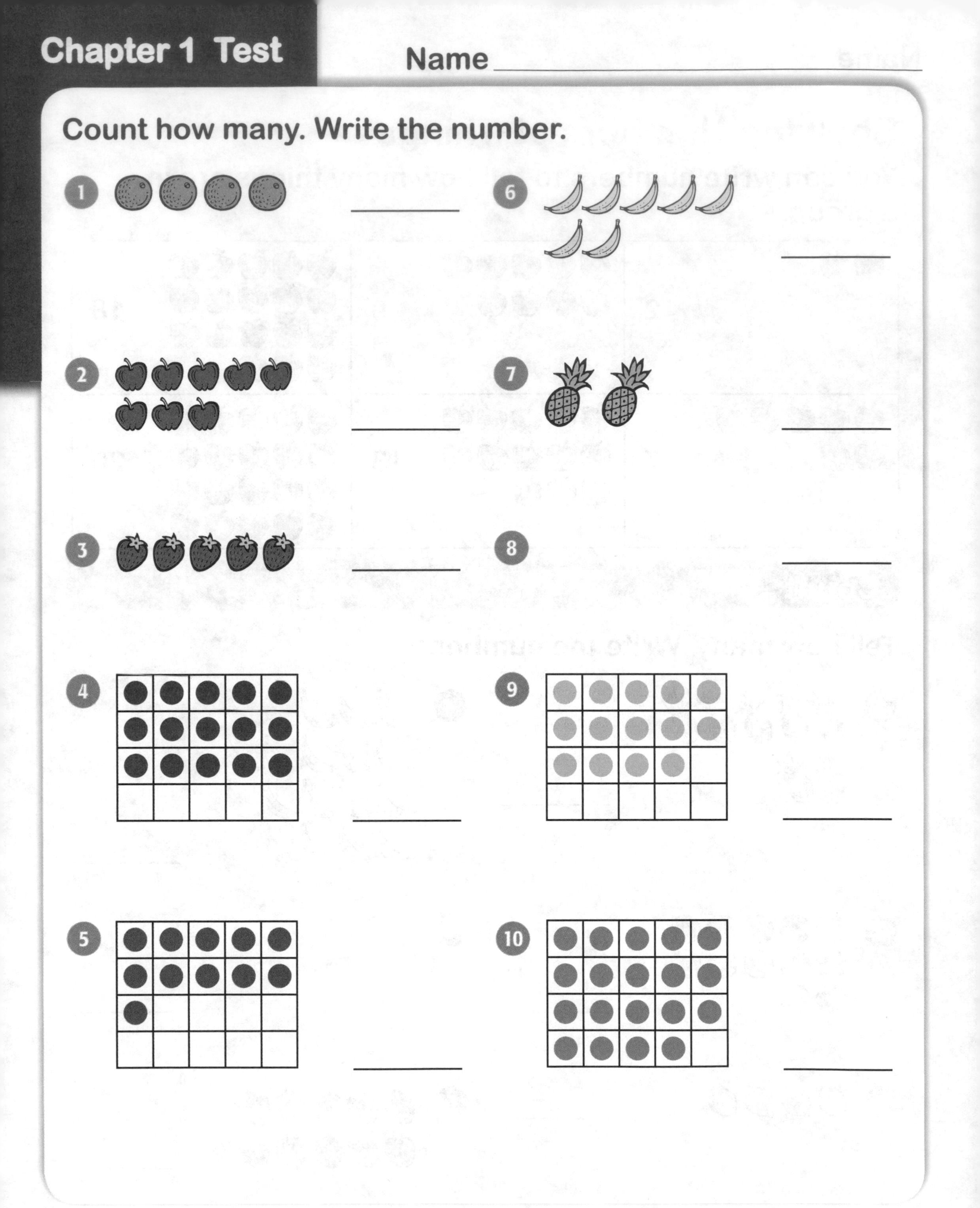

Count how many. Write the number.

1 ______

2 ______

3 ______

4 ______

5 ______

6 ______

7 ______

8 ______

9 ______

10 ______

Name ____________________

Count how many. Write the number.

11 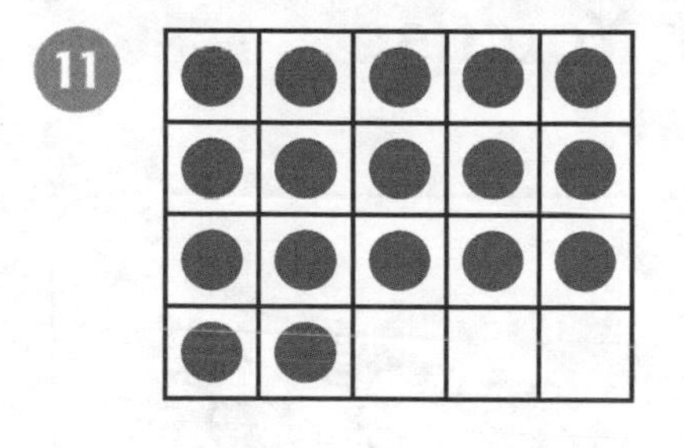 ______

12 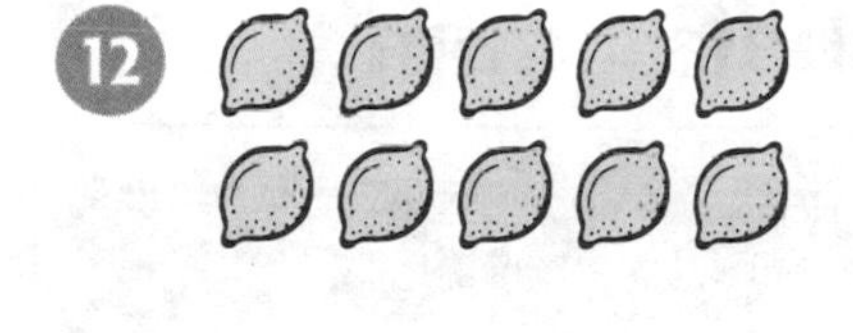 ______

13  ______

14 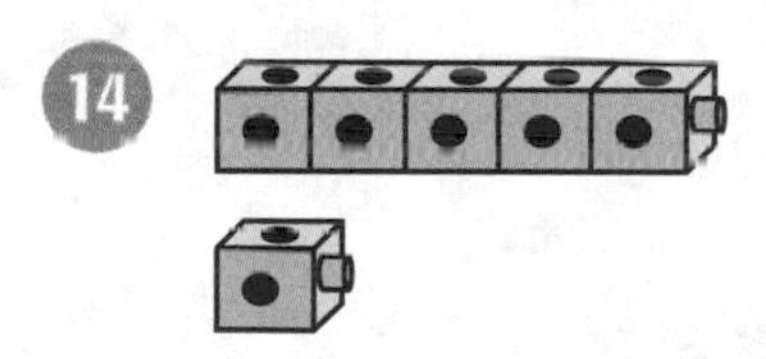 ______

15  ______

16 ______

17 ______

18  ______

19 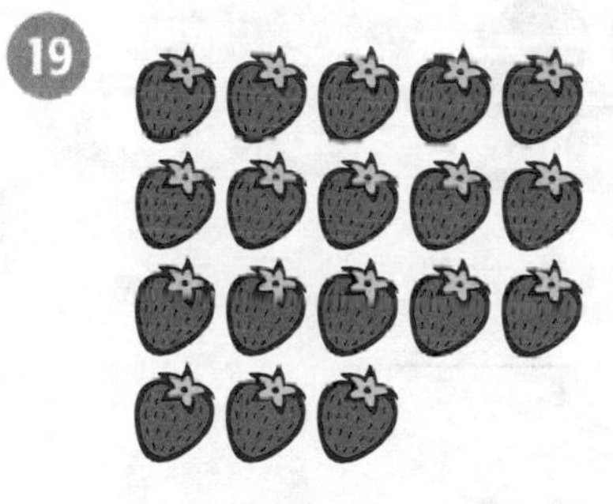 ______

20 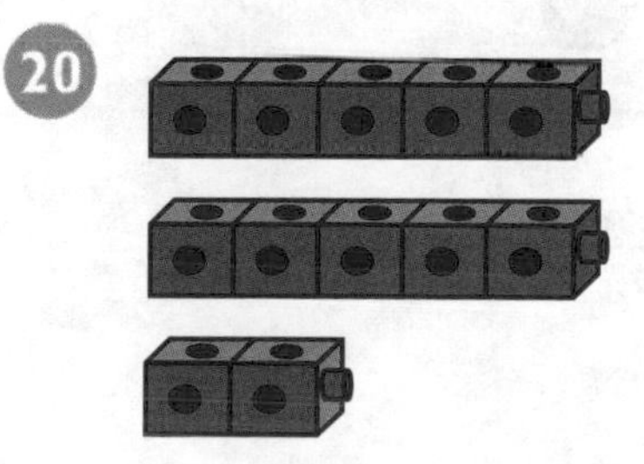 ______

Name ____________________

Addition Facts Through 6

When you add, you group things together. You use a number sentence to show the sum.

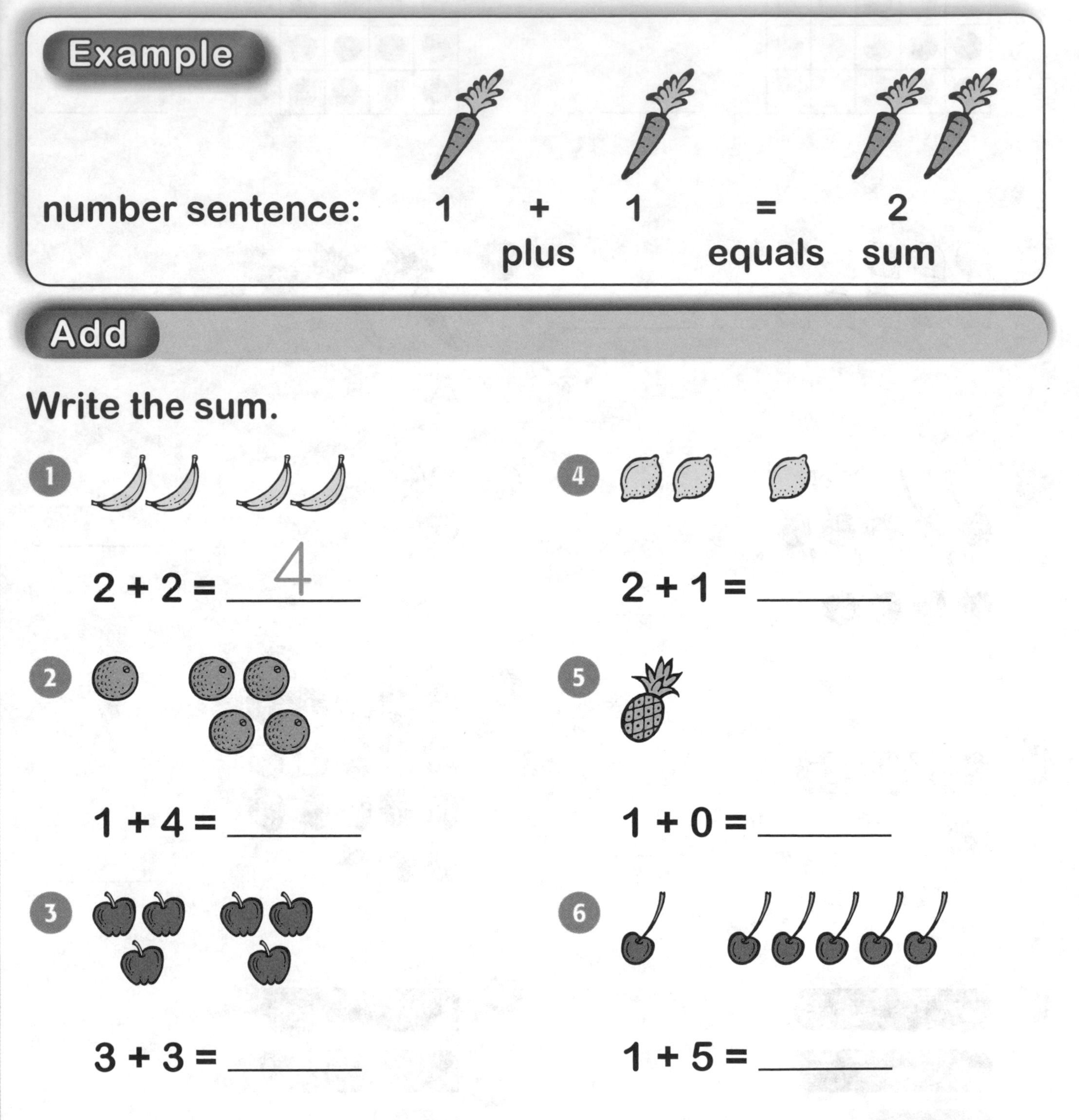

Example

number sentence: 1 + 1 = 2

+ plus, = equals, 2 sum

Add

Write the sum.

1. 2 + 2 = 4
2. 1 + 4 = ______
3. 3 + 3 = ______
4. 2 + 1 = ______
5. 1 + 0 = ______
6. 1 + 5 = ______

Name ______________________

Addition Facts Through 12

You can add by counting how many in each group. Then count how many in all.

Example

Count each group.

5 + 3

Count how many in all.

5 + 3 = 8 in all

Add

Write the sum.

1. 7 + 3 = 10

2. 4 + 5 = ______

3. 5 + 6 = ______

4. 8 + 4 = ______

5. 6 + 1 = ______

6. 4 + 4 = ______

Name ____________________

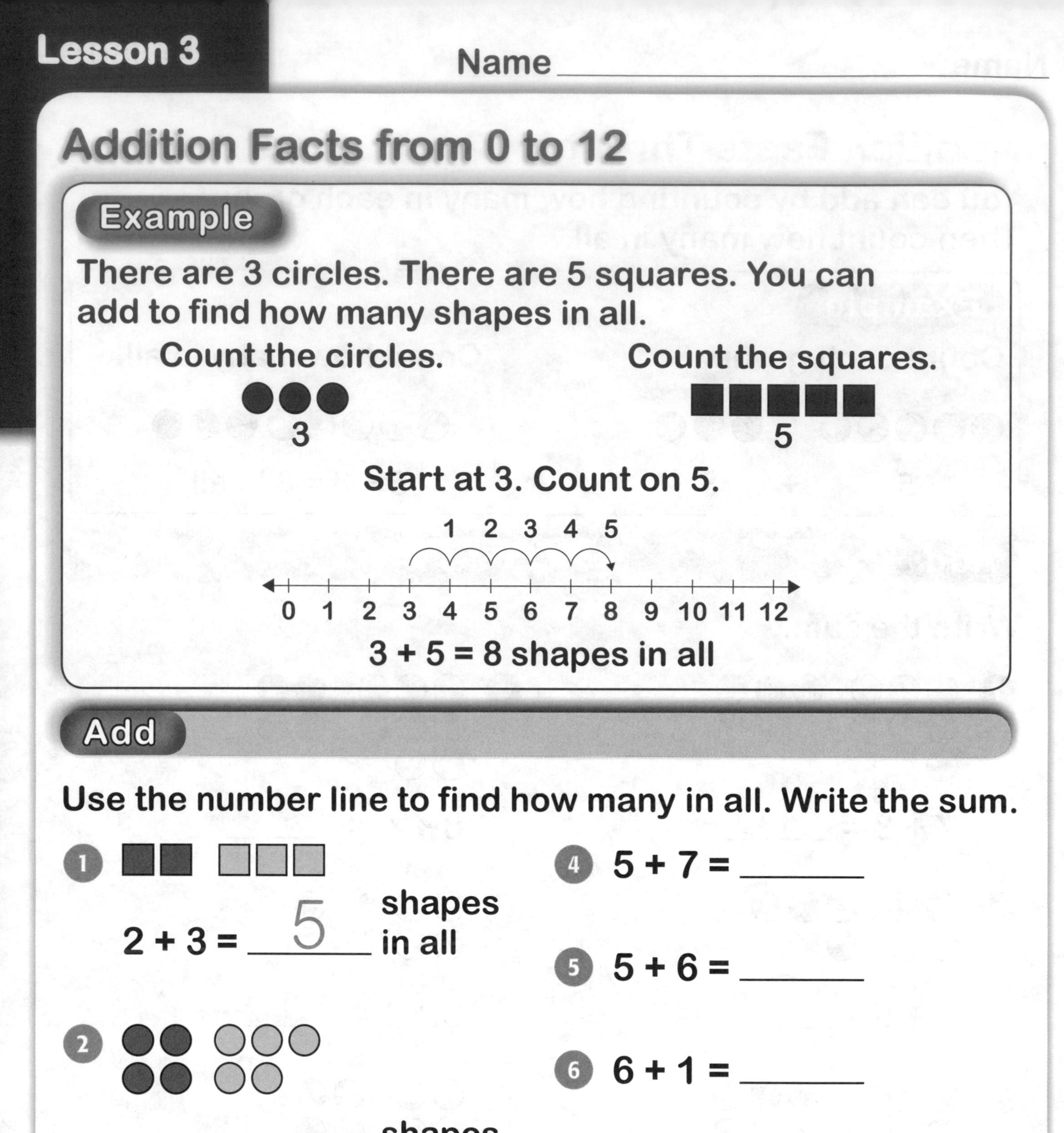

Addition Facts from 0 to 12

Example

There are 3 circles. There are 5 squares. You can add to find how many shapes in all.

Count the circles. ●●● 3

Count the squares. ■■■■■ 5

Start at 3. Count on 5.

1 2 3 4 5

0 1 2 3 4 5 6 7 8 9 10 11 12

3 + 5 = 8 shapes in all

Add

Use the number line to find how many in all. Write the sum.

1. ■■ ■■■

   2 + 3 = 5 shapes in all

2. ●● ●●● / ●● ●●

   4 + 5 = ______ shapes in all

3. 1 + 0 = ______

4. 5 + 7 = ______

5. 5 + 6 = ______

6. 6 + 1 = ______

7. 4 + 2 = ______

8. 2 + 1 = ______

Name ____________________

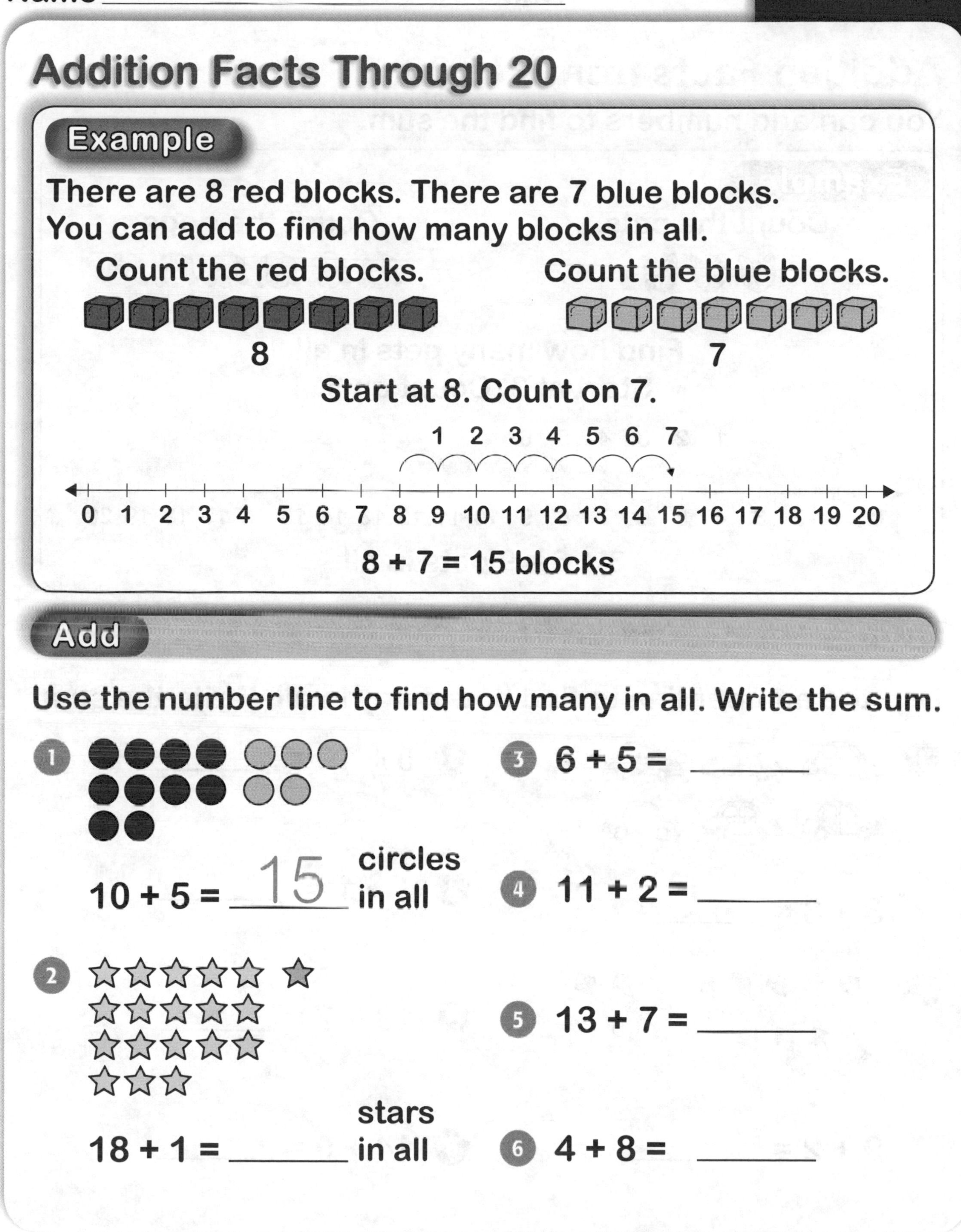

Addition Facts Through 20

Example

There are 8 red blocks. There are 7 blue blocks.
You can add to find how many blocks in all.

Count the red blocks. — 8

Count the blue blocks. — 7

Start at 8. Count on 7.

1 2 3 4 5 6 7

0 1 2 3 4 5 6 7 8 9 10 11 12 13 14 15 16 17 18 19 20

8 + 7 = 15 blocks

Add

Use the number line to find how many in all. Write the sum.

1. 10 + 5 = 15 circles in all

2. 18 + 1 = ______ stars in all

3. 6 + 5 = ______

4. 11 + 2 = ______

5. 13 + 7 = ______

6. 4 + 8 = ______

Name ______________________

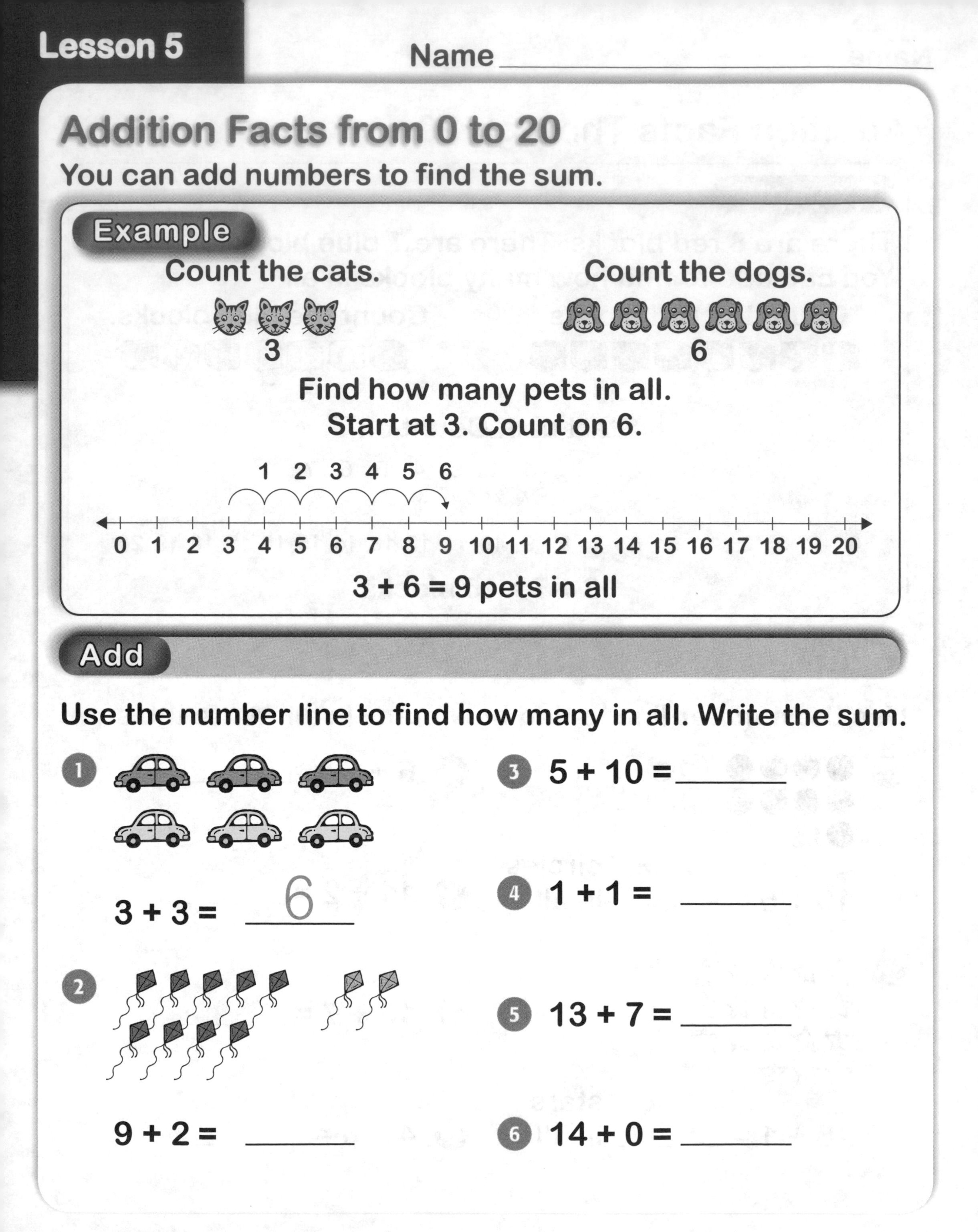

Addition Facts from 0 to 20

You can add numbers to find the sum.

Example

Count the cats. 3

Count the dogs. 6

Find how many pets in all.
Start at 3. Count on 6.

1 2 3 4 5 6

0 1 2 3 4 5 6 7 8 9 10 11 12 13 14 15 16 17 18 19 20

3 + 6 = 9 pets in all

Add

Use the number line to find how many in all. Write the sum.

1. 3 + 3 = 6

2. 9 + 2 = ______

3. 5 + 10 = ______

4. 1 + 1 = ______

5. 13 + 7 = ______

6. 14 + 0 = ______

Name ____________________

Addition Word Problems

You can add to solve word problems.

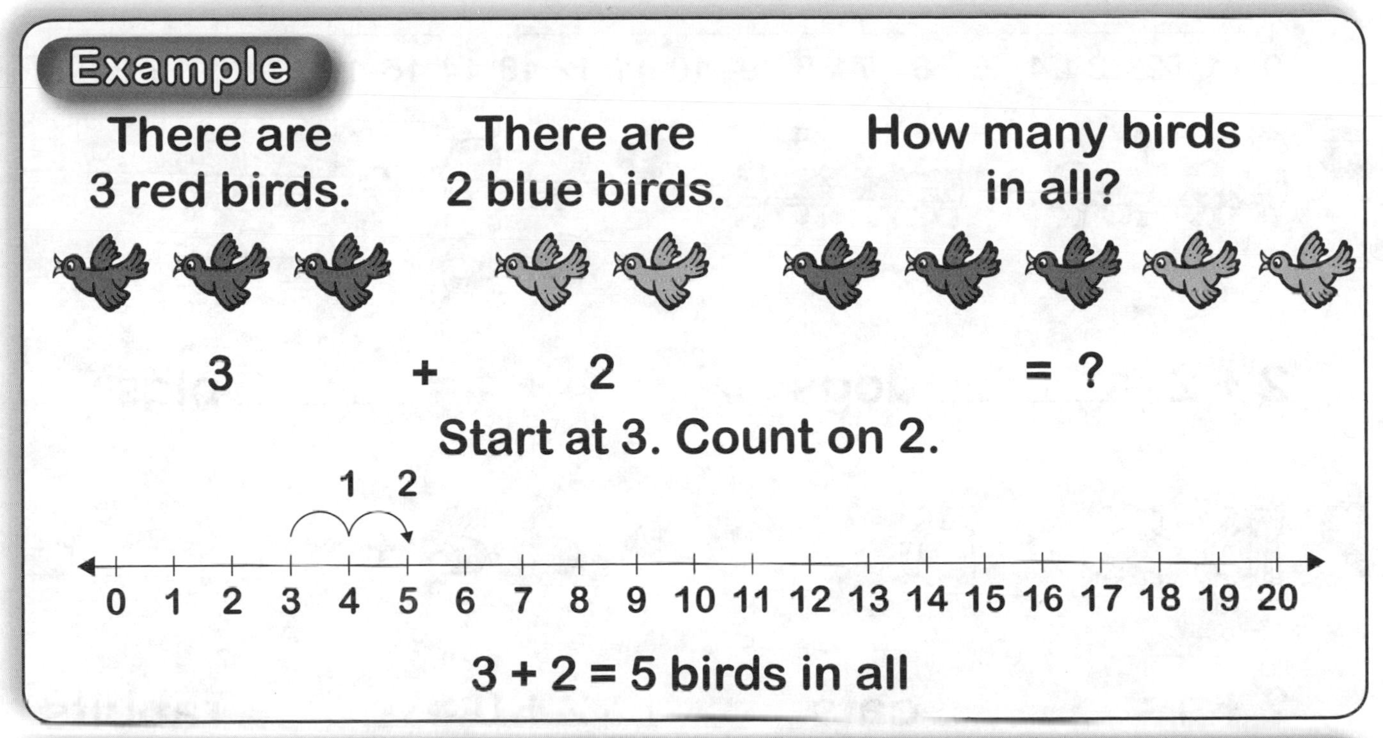

Solve

Add to solve each problem. Use the number line to help.

1. Sam has 7 crayons.
   Ana has 2 crayons.
   How many crayons in all?

   7 + 2 = __9__ crayons

2. There are 4 apples in a bowl.
   There are 2 apples in a bag.
   How many apples in all?

   4 + 2 = ______ apples

3. Dan has 6 books.
   Lily has 8 books.
   How many books are there?

   6 + 8 = ______ books

4. I see 6 dogs and 4 cats.
   How many pets do I see?

   6 + 4 = ______ pets

Name ______________________

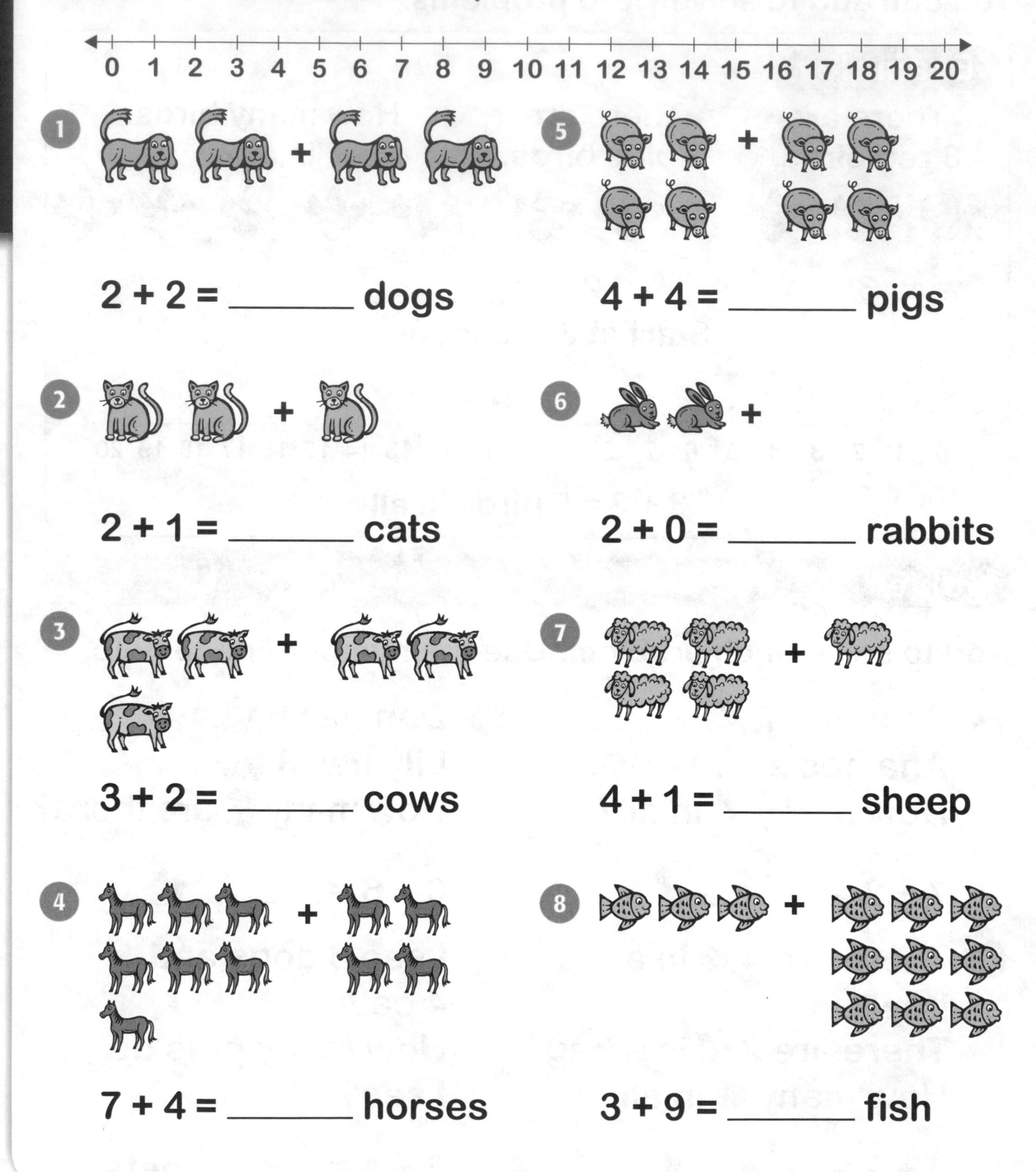

Add to find how many in all. Write the sum.
Use the number line to help.

0 1 2 3 4 5 6 7 8 9 10 11 12 13 14 15 16 17 18 19 20

1. 2 + 2 = ______ dogs

2. 2 + 1 = ______ cats

3. 3 + 2 = ______ cows

4. 7 + 4 = ______ horses

5. 4 + 4 = ______ pigs

6. 2 + 0 = ______ rabbits

7. 4 + 1 = ______ sheep

8. 3 + 9 = ______ fish

Name ____________________

Add. Write the sum.

9. $5 + 5 =$ ______

14. $3 + 5 =$ ______

10. $2 + 7 =$ ______

15. $14 + 5 =$ ______

11. $1 + 1 =$ ______

16. $8 + 5 =$ ______

12. $6 + 3 =$ ______

17. $9 + 8 =$ ______

13. $1 + 11 =$ ______

18. $0 + 14 =$ ______

Solve. Write the sum. Use the number line to help.

0 1 2 3 4 5 6 7 8 9 10 11 12 13 14 15 16 17 18 19 20

19. There are 5 pups playing.
There are 2 pups sleeping.
How many pups are there?

$5 + 2 =$ ______ pups

20. Ed has 7 pens.
His mom has 5 pens.
How many do they have in all?

$7 + 5 =$ ______ pens

Name ____________________

Subtraction Facts Through 6

When you subtract, you take away from a group. You use a number sentence to show the difference.

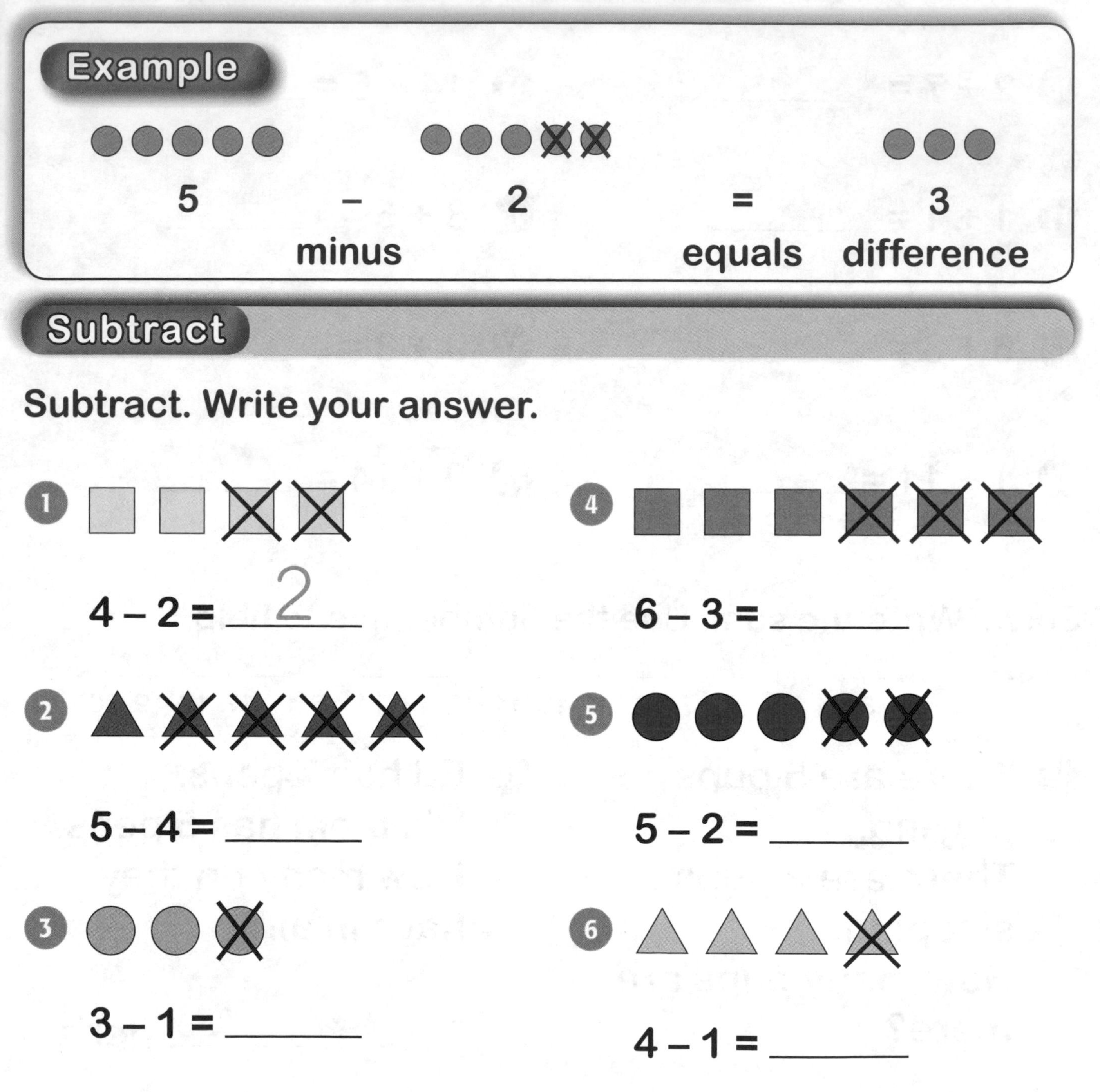

Example

5 – 2 = 3

minus equals difference

Subtract

Subtract. Write your answer.

1. 4 – 2 = 2
2. 5 – 4 = ______
3. 3 – 1 = ______
4. 6 – 3 = ______
5. 5 – 2 = ______
6. 4 – 1 = ______

Name ______________________

Subtraction Facts Through 12

You can count back on a number line to subtract.

Example

$12 - 5 = ?$

Start at 12. Count back 5.

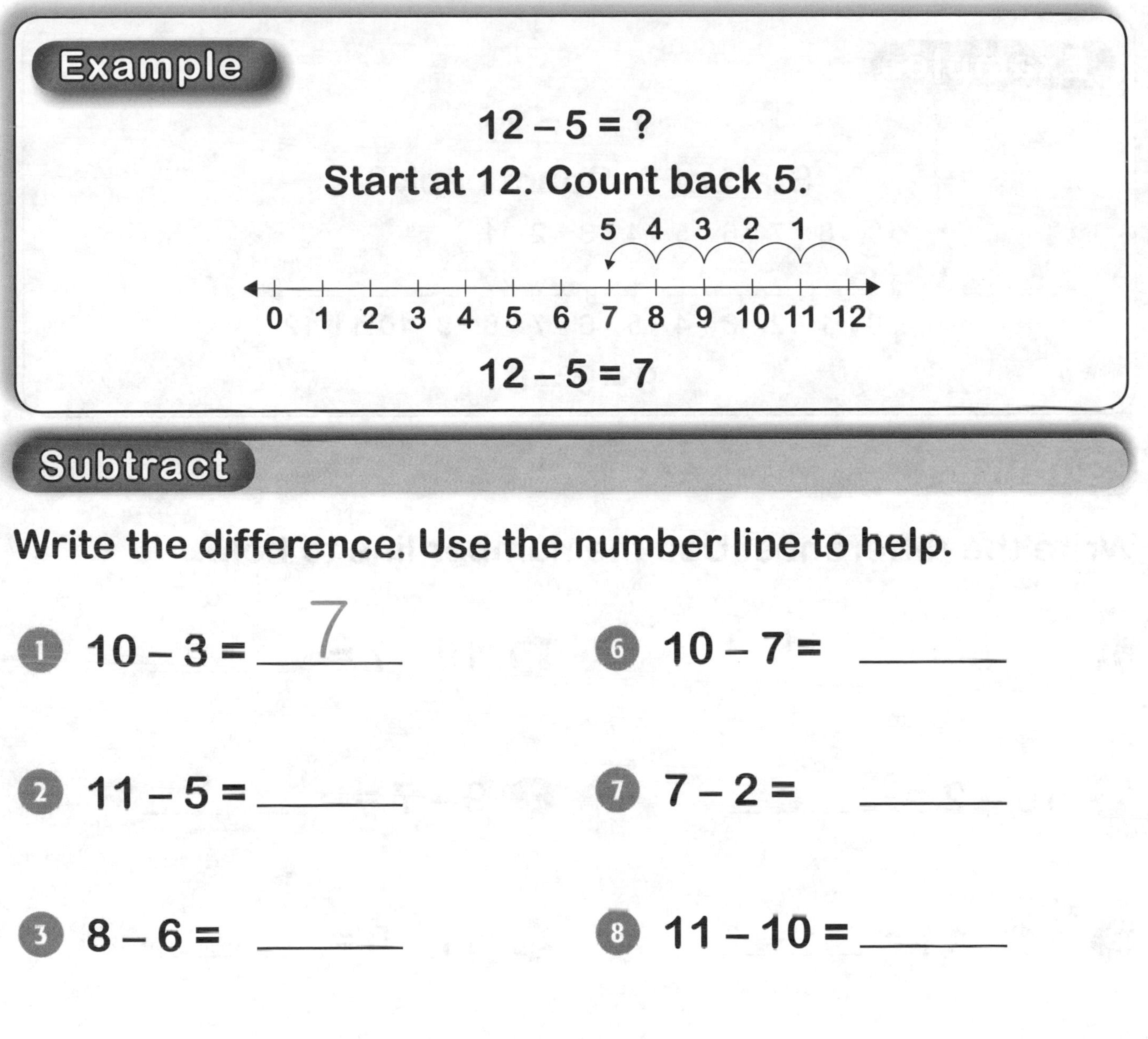

$12 - 5 = 7$

Subtract

Write the difference. Use the number line to help.

1. $10 - 3 =$ 7
2. $11 - 5 =$ ______
3. $8 - 6 =$ ______
4. $12 - 1 =$ ______
5. $10 - 5 =$ ______
6. $10 - 7 =$ ______
7. $7 - 2 =$ ______
8. $11 - 10 =$ ______
9. $9 - 3 =$ ______
10. $12 - 8 =$ ______

Name ______________________

Subtraction Facts from 0 to 12

You can count back on a number line to subtract.

Example

$9 - 9 = ?$

Start at 9. Count back 9.

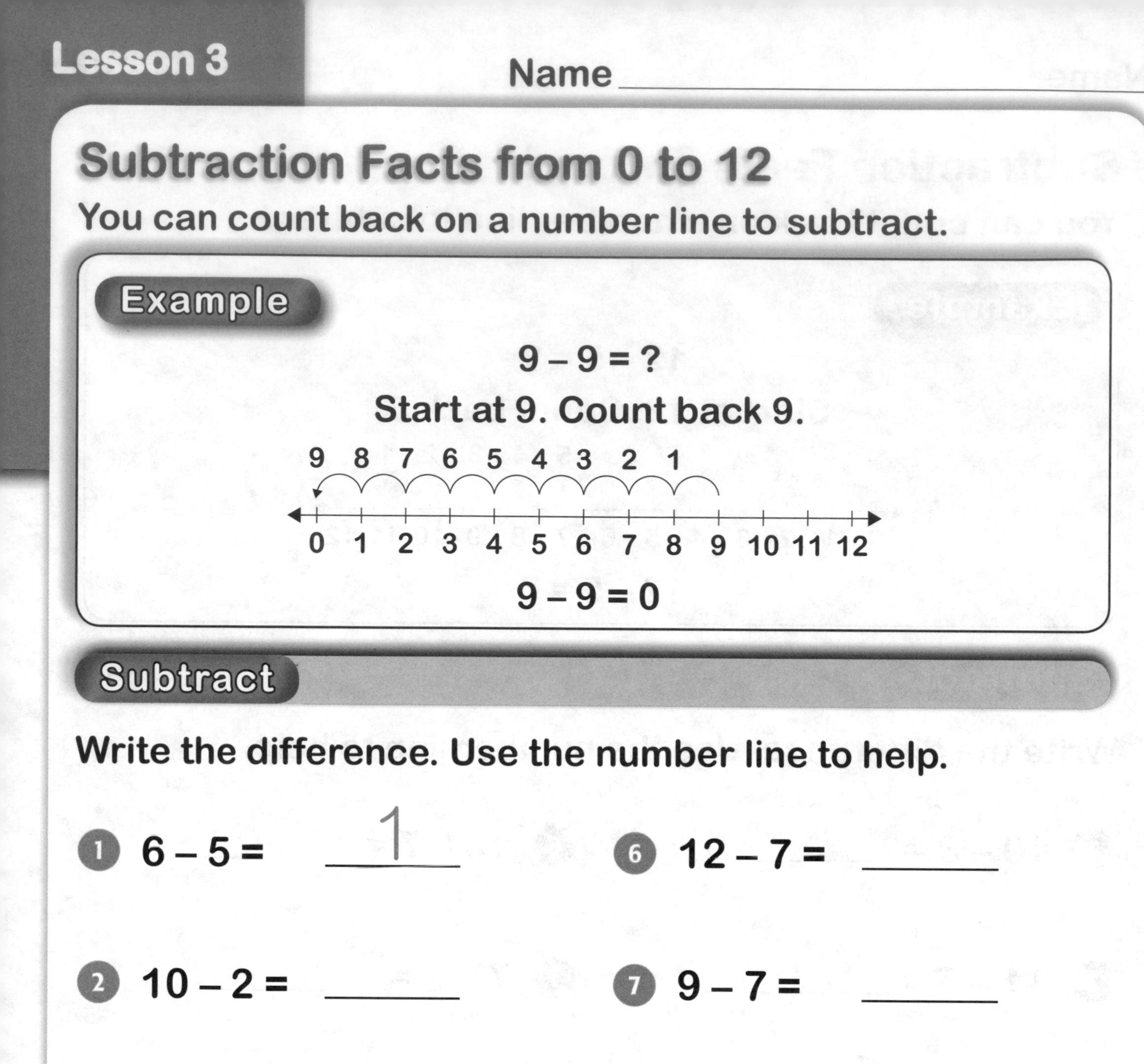

$9 - 9 = 0$

Subtract

Write the difference. Use the number line to help.

1. $6 - 5 =$ __1__
2. $10 - 2 =$ ______
3. $11 - 11 =$ ______
4. $7 - 4 =$ ______
5. $7 - 1 =$ ______
6. $12 - 7 =$ ______
7. $9 - 7 =$ ______
8. $11 - 0 =$ ______
9. $12 - 12 =$ ______
10. $8 - 4 =$ ______

Name ________________________

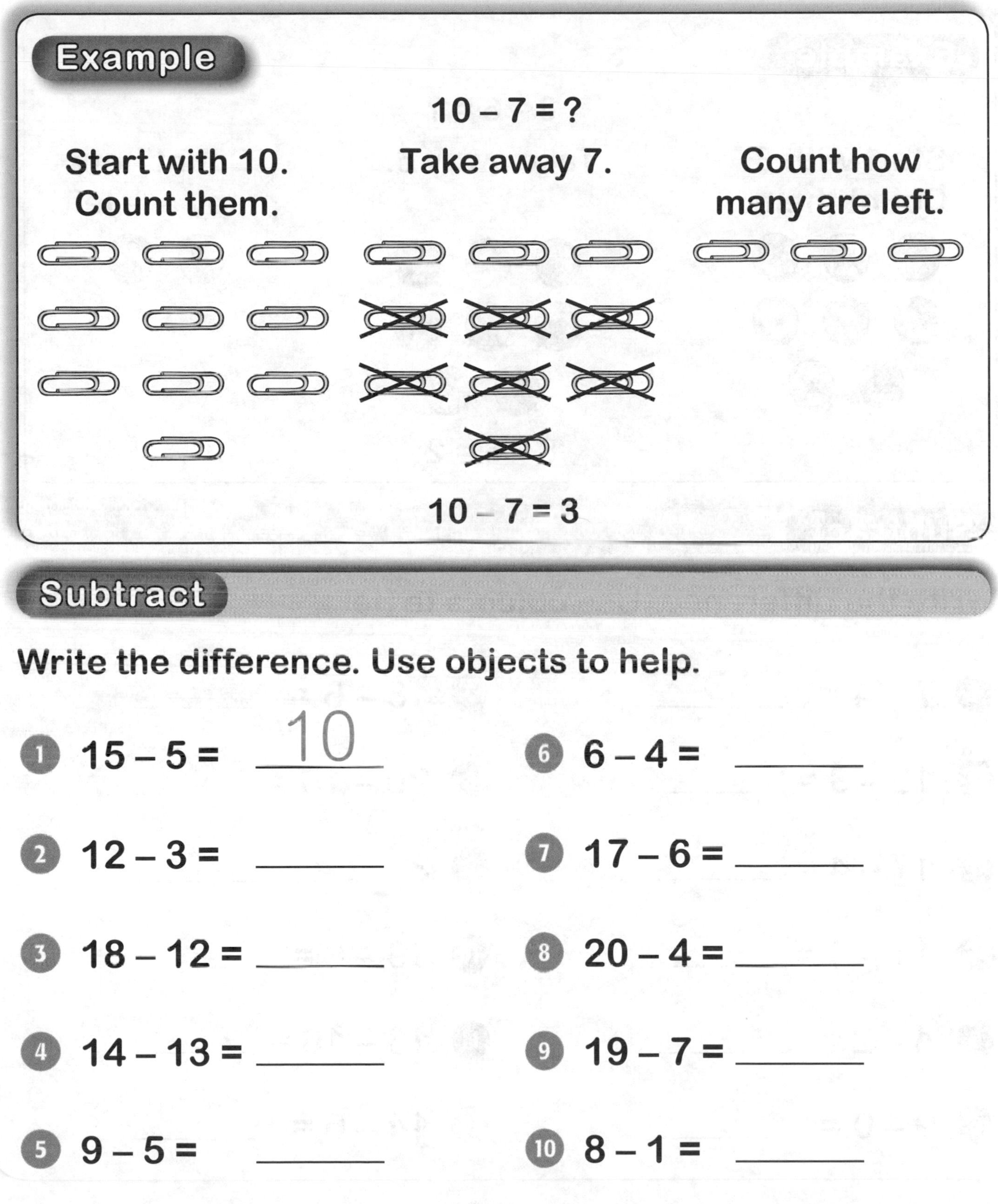

Subtraction Facts Through 20

You can use objects to help you subtract.

Example

10 – 7 = ?

Start with 10. Count them.	Take away 7.	Count how many are left.

10 – 7 = 3

Subtract

Write the difference. Use objects to help.

1. 15 – 5 = 10
2. 12 – 3 = ______
3. 18 – 12 = ______
4. 14 – 13 = ______
5. 9 – 5 = ______
6. 6 – 4 = ______
7. 17 – 6 = ______
8. 20 – 4 = ______
9. 19 – 7 = ______
10. 8 – 1 = ______

Name ____________________

Subtraction Facts from 0 to 20

You can use small objects to help you subtract.

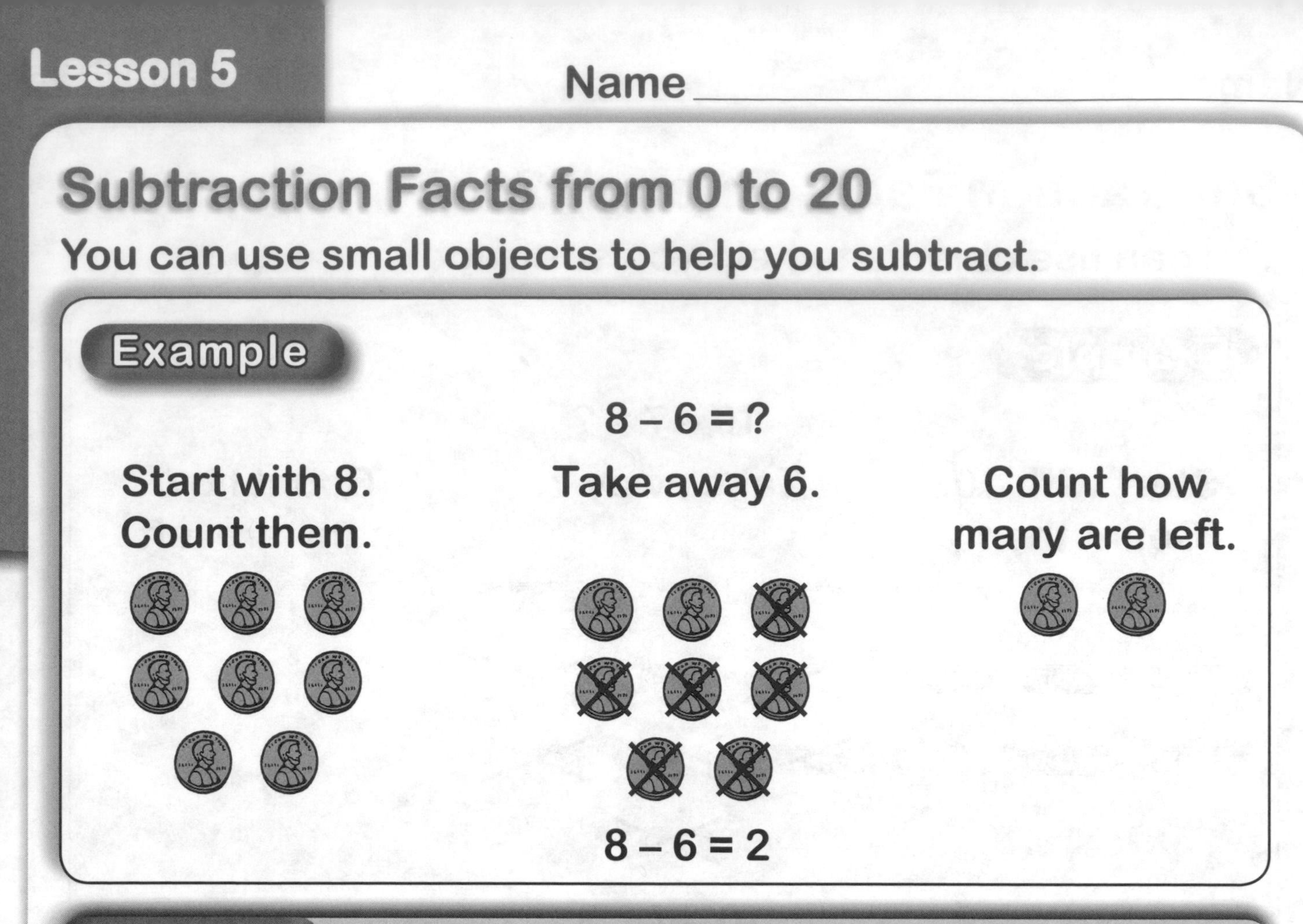

Subtract

Write the difference. Use objects to help.

1. 7 – 4 = 3
2. 13 – 3 = ______
3. 17 – 4 = ______
4. 11 – 4 = ______
5. 4 – 2 = ______
6. 9 – 0 = ______
7. 18 – 6 = ______
8. 20 – 15 = ______
9. 3 – 2 = ______
10. 19 – 8 = ______
11. 16 – 16 = ______
12. 14 – 6 = ______

Name ______________________

Subtraction Word Problems

Subtract to solve problems.

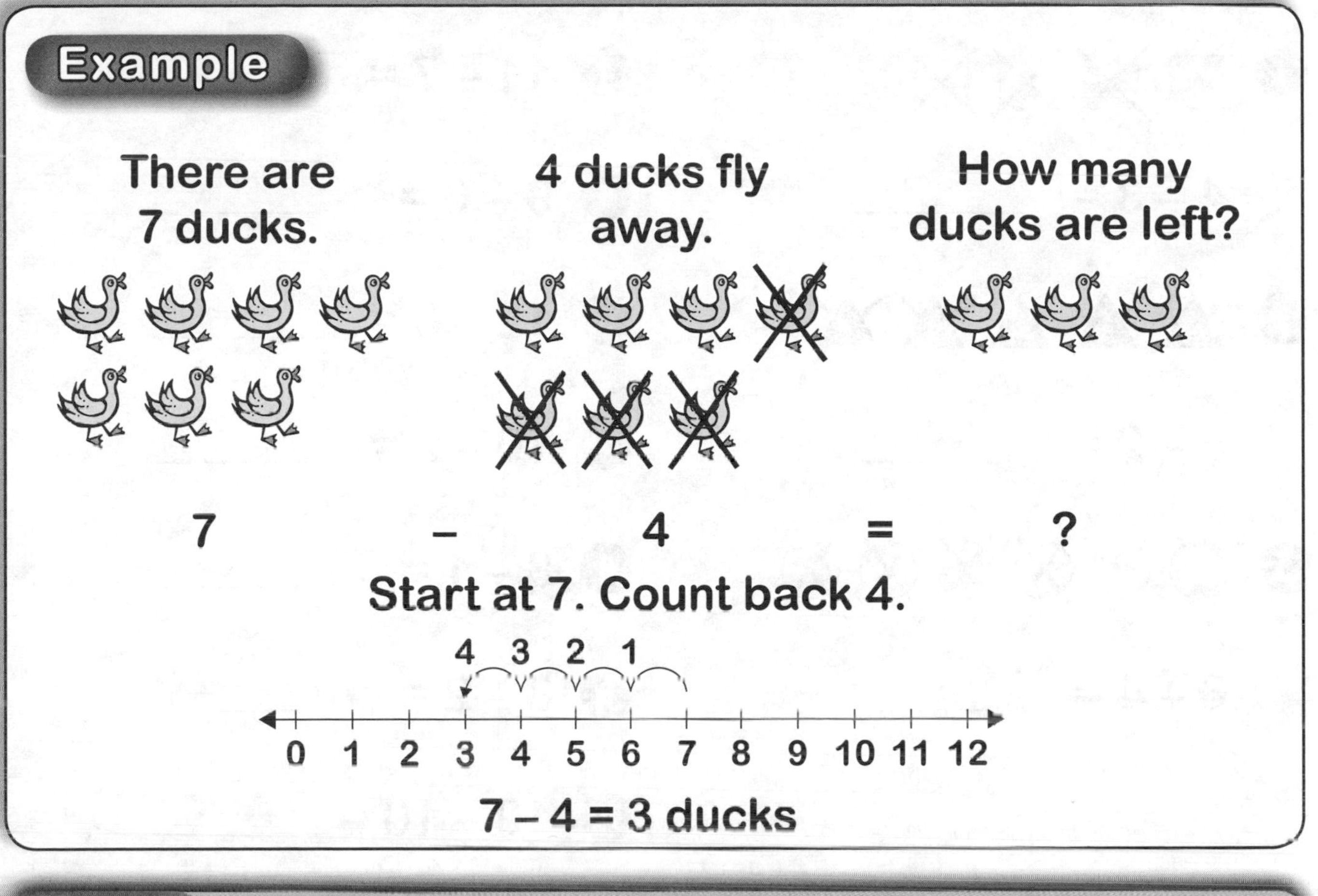

Solve

Subtract to solve the problems. Write the difference.

1. Sue has 8 in a bag.
   Sue gives 3 away.
   How many does Sue have now?

   8 – 3 = __5__ apples

2. There are 17 in the field.
   15 go into the barn.
   How many are in the field now?

   17 – 15 = ______ horses

Name ____________________

Subtract. Write the difference. Use the number line to help.

0 1 2 3 4 5 6 7 8 9 10 11 12 13 14 15 16 17 18 19 20

1. $4 - 3 =$ ______

2. $5 - 2 =$ ______

3. $6 - 4 =$ ______

4. $11 - 7 =$ ______

5. $9 - 6 =$ ______

6. $12 - 6 =$ ______

7. $10 - 2 =$ ______

8. $8 - 1 =$ ______

9. $3 - 2 =$ ______

10. $10 - 10 =$ ______

11. $11 - 6 =$ ______

Subtract. Write the difference. Use small objects to help.

12. $7 - 3 =$ ______

13. $16 - 5 =$ ______

14. $19 - 9 =$ ______

15. $14 - 6 =$ ______

16. $20 - 17 =$ ______

17. $15 - 15 =$ ______

18. $6 - 0 =$ ______

19. $12 - 7 =$ ______

Name ______________________________

Solve the problems. Write the difference. Use the number line to help.

0 1 2 3 4 5 6 7 8 9 10 11 12 13 14 15 16 17 18 19 20

20. Carlos has 19 .
Carlos gives 5 to Tom.
How many does Carlos have now?

$19 - 5 =$ ________

21. Tina has 18 .
Tina gives 2 to Dad.
How many does Tina have left?

$18 - 2 =$ ________

22. A pet store has 20 .
A man buys 7 .
How many are left?

$20 - 7 =$ ________

23. A bag has 19 .
11 fall out.
How many are in the bag now?

$19 - 11 =$ ________

24. Ms. Lee has 13 .
She gives away 8 .
How many does she have left?

$13 - 8 =$ ________

25. 7 dogs are in the park.
4 dogs go home.
How many dogs are in the park now?

$7 - 4 =$ ________

26. Kay sees 10 ants.
5 ants go away.
How many ants are left?

$10 - 5 =$ ________ ants

Name ____________________

Counting Forward

You can start counting with any number.

0	zero
1	one
2	two
3	three
4	four
5	five
6	six
7	seven
8	eight
9	nine
10	ten

11	eleven
12	twelve
13	thirteen
14	fourteen
15	fifteen
16	sixteen
17	seventeen
18	eighteen
19	nineteen
20	twenty

Start with 6. Count to 10.

6 7 8 9 10

Count

Write the numbers.

1. 0 1 2 3 4
2. 4 5 ____ ____ ____
3. 6 7 ____ ____ ____
4. 14 15 ____ ____ ____
5. 11 12 ____ ____ ____
6. 8 9 ____ ____ ____
7. 16 17 ____ ____ ____
8. 9 10 ____ ____ ____

Name ______________________________

Counting from 0 to 50

You can count from 0 to 50.

0

1	2	3	4	5	6	7	8	9	10
11	12	13	14	15	16	17	18	19	20
21	22	23	24	25	26	27	28	29	30
31	32	33	34	35	36	37	38	39	40
41	42	43	44	45	46	47	48	49	50

Count

Write the missing numbers. Then count aloud.

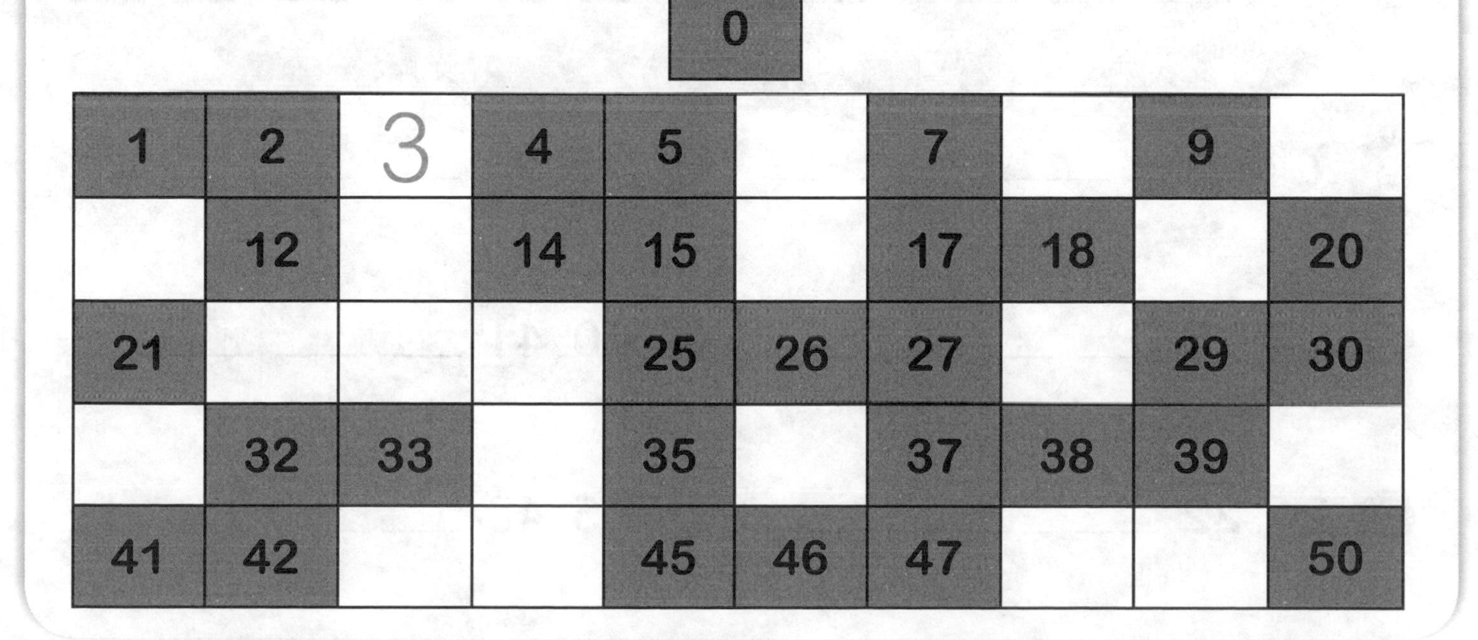

0

1	2	3	4	5		7		9	
	12		14	15		17	18		20
21				25	26	27		29	30
	32	33		35		37	38	39	
41	42			45	46	47			50

Lesson 3

Name ____________________

Writing Numbers from 0 to 50

You can write numbers from 0 to 50.

				0					
1	2	3	4	5	6	7	8	9	10
11	12	13	14	15	16	17	18	19	20
21	22	23	24	25	26	27	28	29	30
31	32	33	34	35	36	37	38	39	40
41	42	43	44	45	46	47	48	49	50

Count

Write the numbers. Use the chart to help.

1. 0 1 2 3 4 5 6
2. 7 8 ___ ___ ___ ___
3. 14 15 ___ ___ ___ ___
4. 21 22 ___ ___ ___ ___
5. 28 29 ___ ___ ___ ___
6. 34 35 ___ ___ ___ ___
7. 40 41 ___ ___ ___ ___
8. 45 46 ___ ___ ___ ___

Name ____________________

Counting Forward from Any Number

You can start counting with any number.

Example

Start with 14. Count to 18.

14 15 16 17 18

Count

Write the numbers.

1. 6 7 8 9 10
2. 22 23 ___ ___ ___
3. 0 1 ___ ___ ___
4. Start with 2.
   ___ ___ ___ ___
5. Start with 33.
   ___ ___ ___ ___
6. 18 19 ___ ___ ___
7. 43 44 ___ ___ ___
8. 17 18 ___ ___ ___
9. Start with 25.
   ___ ___ ___ ___
10. Start with 47.
    ___ ___ ___ ___

Name ____________________

Tell what number comes next. Write the number.

1. 0 ____
2. 5 ____
3. 9 ____
4. 3 ____
5. 15 ____
6. 18 ____
7. 7 ____
8. 11 ____

9. **Write the missing numbers.**

0

	2	3	4			7		9	10
11		13				17	18	19	20
21	22		24	25	26	27		29	
	32	33		35	36	37			40
41	42		44	45			48	49	

Name ____________________

Count. Write the numbers.

10. 0 1 ___ ___ ___ ___

11. 13 14 ___ ___ ___ ___

12. 34 35 ___ ___ ___ ___

13. 27 28 ___ ___ ___ ___

14. 39 40 ___ ___ ___ ___

15. 18 19 ___ ___ ___ ___

16. 6 7 ___ ___ ___ ___

17. 45 46 ___ ___ ___ ___

18. Start with 7.

___ ___ ___ ___

19. Start with 19.

___ ___ ___ ___

20. Start with 30.

___ ___ ___ ___

21. Start with 21.

___ ___ ___ ___

22. Start with 42.

___ ___ ___ ___

23. Start with 16.

___ ___ ___ ___

24. Start with 47.

___ ___ ___ ___

25. Start with 3.

___ ___ ___ ___

Name ______________________

Counting from 50 to 75

You can count past 50. You start at 50.
Then you count on.

50	fifty	60	sixty	70	seventy
51	fifty-one	61	sixty-one	71	seventy-one
52	fifty-two	62	sixty-two	72	seventy-two
53	fifty-three	63	sixty-three	73	seventy-three
54	fifty-four	64	sixty-four	74	seventy-four
55	fifty-five	65	sixty-five	75	seventy-five
56	fifty-six	66	sixty-six		
57	fifty-seven	67	sixty-seven		
58	fifty-eight	68	sixty-eight		
59	fifty-nine	69	sixty-nine		

Count

Read the number shown. Count on. Write the numbers.

1. 50 51 52 53
2. 54 ______ ______ ______
3. 61 ______ ______ ______
4. 70 ______ ______ ______
5. 53 ______ ______ ______
6. 58 ______ ______ ______
7. 66 ______ ______ ______
8. 72 ______ ______ ______

Name ____________________

Counting from 76 to 100

You can count past 75. You can start at 76.
Then you count on.

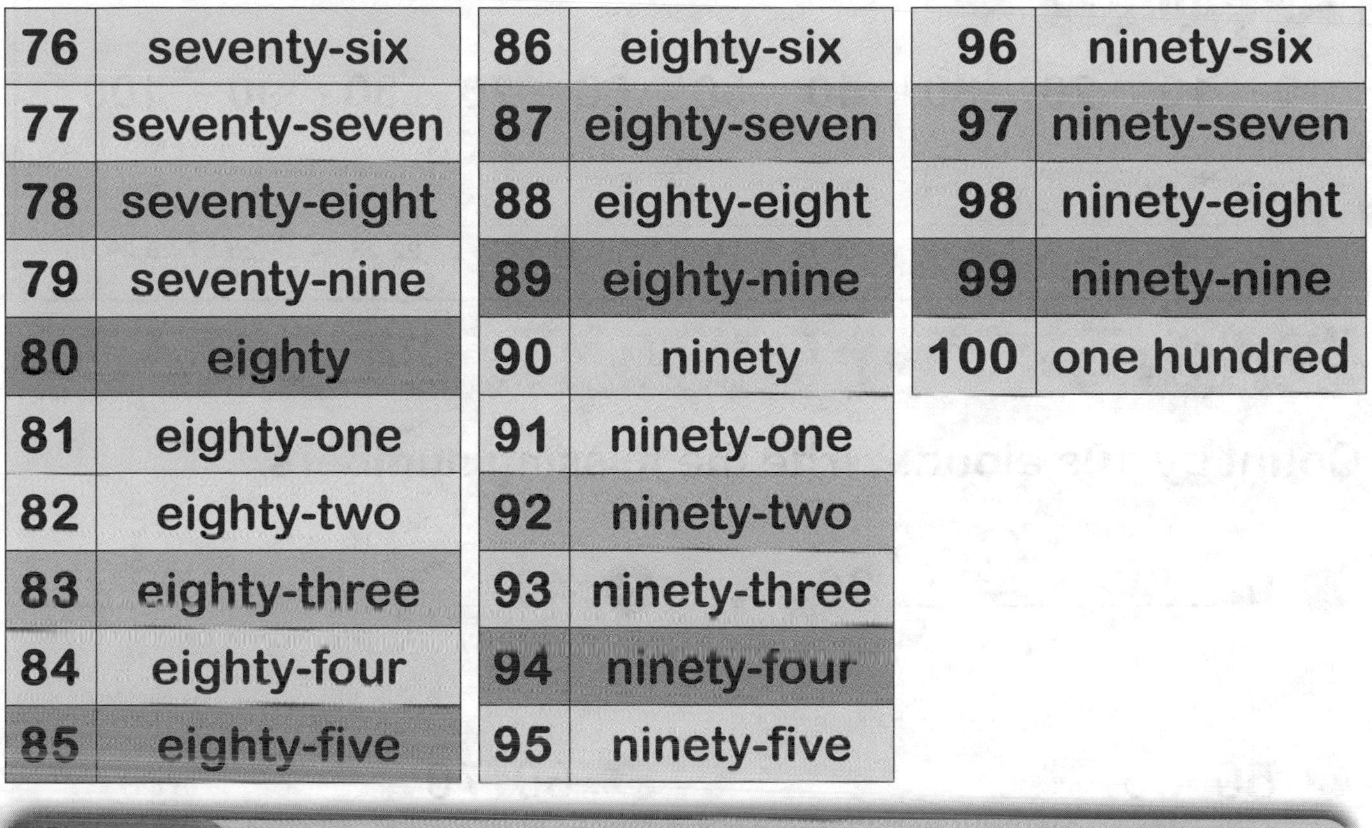

76	seventy-six	86	eighty-six	96	ninety-six
77	seventy-seven	87	eighty-seven	97	ninety-seven
78	seventy-eight	88	eighty-eight	98	ninety-eight
79	seventy-nine	89	eighty-nine	99	ninety-nine
80	eighty	90	ninety	100	one hundred
81	eighty-one	91	ninety-one		
82	eighty-two	92	ninety-two		
83	eighty-three	93	ninety-three		
84	eighty-four	94	ninety-four		
85	eighty-five	95	ninety-five		

Count

Read the number shown. Count on. Write the numbers.

1. 76 77 78 79
2. 82 ____ ____ ____
3. 97 ____ ____ ____
4. 78 ____ ____ ____
5. 90 ____ ____ ____
6. 77 ____ ____ ____
7. 88 ____ ____ ____
8. 94 ____ ____ ____

Name ____________________

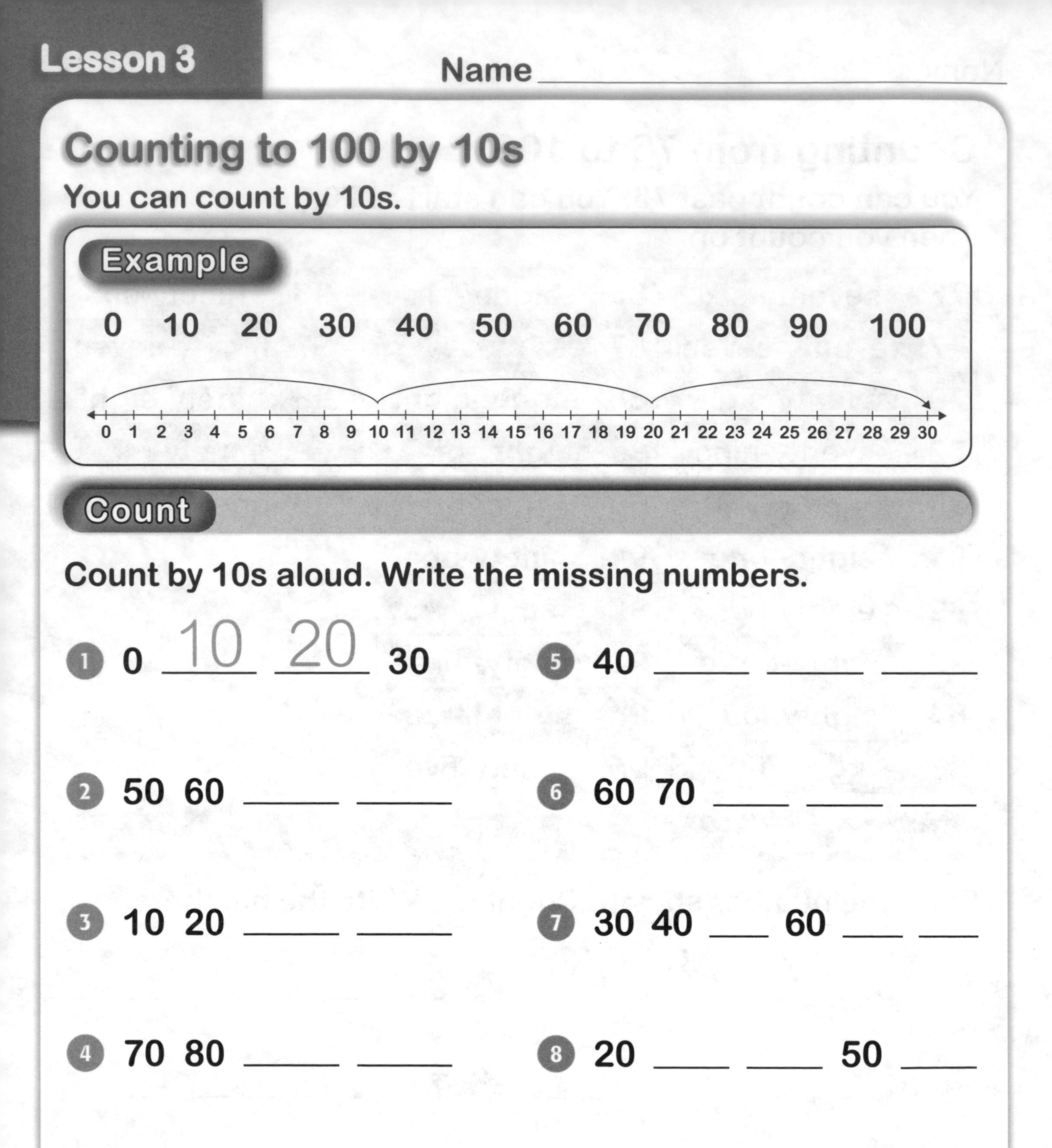

Counting to 100 by 10s

You can count by 10s.

Example

0 10 20 30 40 50 60 70 80 90 100

Count

Count by 10s aloud. Write the missing numbers.

1. 0 10 20 30
2. 50 60 ____ ____
3. 10 20 ____ ____
4. 70 80 ____ ____
5. 40 ____ ____ ____
6. 60 70 ____ ____ ____
7. 30 40 ____ 60 ____ ____
8. 20 ____ ____ 50 ____

Name ______________________

Writing Numbers to 100

You can count to 100. You can write the numbers.

Count

Begin with 50. Complete the chart. Write the missing numbers.

1	2	3	4	5	6	7	8	9	10
11	12	13	14	15	16	17	18	19	20
21	22	23	24	25	26	27	28	29	30
31	32	33	34	35	36	37	38	39	40
41	42	43	44	45	46	47	48	49	50
51									

Lesson 5

Name ______________________

Counting Forward to 100

You can count from 0 to 100. Start with any number. Then you count on.

Example

Start with 0.

0 1 2 3 4 5 6 7 8 9 10

Start with 52.

52 53 54 55 56 57 58 59 60 61 62

Count

Count aloud. Write the missing numbers.

1. 1 2 3 __4__ __5__ __6__ 7 8 __9__
2. 76 77 ______ 79 ______ ______ 82 ______ 84
3. 91 92 ______ 94 ______ 96 97 ______ ______
4. 50 ______ 52 ______ 54 ______ 56 ______ 58
5. 82 ______ 84 ______ ______ 87 ______ 89 90
6. 67 68 69 ______ ______ ______ ______ 74 75
7. 17 18 ______ ______ 21 22 ______ 24 ______
8. 33 ______ ______ 36 ______ 38 39 ______ 41

Name ______________________

1 Complete the chart. Write the missing numbers.

1	2		4	5	6	7		9	10
11	12	13	14	15		17	18		20
21		23	24		26	27	28	29	30
	32	33	34	35	36		38	39	40
41	42	43	44	45	46	47	48		
51		53		55	56	57	58	59	60
61	62	63	64	65	66	67		69	
	72	73	74		76	77	78	79	80
81	82		84	85		87	88	89	90
91	92	93		95	96		98	99	100

Count on. Write the missing numbers.

2 53 54 ____ ____ ____

3 60 61 ____ ____ ____

4 56 57 ____ ____ ____

5 69 70 ____ ____ ____

6 71 72 ____ ____ ____

7 79 80 ____ ____ ____

8 96 97 ____ ____ ____

9 77 78 ____ ____ ____

10 87 88 ____ ____ ____

11 89 90 ____ ____ ____

Name ____________________

Count by 10s. Write the missing numbers.

12. 10 20 ____ ____ ____

14. 30 40 ____ ____ ____

13. 50 60 ____ ____ ____

15. 60 70 ____ ____ ____

Count. Write the missing numbers.

16. 50 ____ ____ 53

21. 33 34 ____ 36 ____

17. 85 ____ ____ ____

22. 67 68 ____ ____

18. 41 ____ 43 ____

23. 80 ____ 82 ____

19. 39 40 ____ ____

24. 91 ____ ____ ____

20. 96 97 ____ ____

25. 17 ____ ____ 20

Name ______________________________

Counting from 100 to 120 by 1s

You can say and write numbers as you count.

100	one hundred	111	one hundred eleven
101	one hundred one	112	one hundred twelve
102	one hundred two	113	one hundred thirteen
103	one hundred three	114	one hundred fourteen
104	one hundred four	115	one hundred fifteen
105	one hundred five	116	one hundred sixteen
106	one hundred six	117	one hundred seventeen
107	one hundred seven	118	one hundred eighteen
108	one hundred eight	119	one hundred nineteen
109	one hundred nine	120	one hundred twenty
110	one hundred ten		

Count

Write the missing numbers.

1. Count from 100 to 103.

   100 101 102 103

2. Count from 106 to 108.

   ______ ______ ______

3. Count from 111 to 113.

   ______ ______ ______

4. Count from 112 to 114.

   ______ ______ ______

5. Count from 105 to 107.

   ______ ______ ______

6. Count from 102 to 104.

   ______ ______ ______

Lesson 2

Name ______________________

Counting from 0 to 120 by 10s

You can count to 120 by 10s.

0	zero
10	ten
20	twenty
30	thirty
40	forty
50	fifty
60	sixty

70	seventy
80	eighty
90	ninety
100	one hundred
110	one hundred ten
120	one hundred twenty

Count

Count by tens out loud. Write the missing numbers.

1. Count from 10 to 40 by 10s.

   10 20 30 40

2. Count from 80 to 110 by 10s.

   80 90 ______ ______

3. Count from 20 to 50 by 10s.

   ______ ______ ______ ______

4. Count from 80 to 110 by 10s.

   ______ ______ ______ ______

5. Count from 0 to 30 by 10s.

   ______ ______ ______ ______

6. Count from 50 to 80 by 10s.

   ______ ______ ______ ______

Name ______________________

Counting from 0 to 100 by 5s

You can count by 5s.

0	zero
5	five
10	ten
15	fifteen
20	twenty
25	twenty-five
30	thirty
35	thirty-five
40	forty
45	forty-five
50	fifty

55	fifty-five
60	sixty
65	sixty-five
70	seventy
75	seventy-five
80	eighty
85	eighty-five
90	ninety
95	ninety-five
100	one hundred

Count

Count by 5s. Write the numbers.

1. Count from 5 to 20 by 5s.

   5 10 15 20

2. Count from 30 to 45 by 5s.

   30 35 ______ ______

3. Count from 55 to 70 by 5s.

   55 60 ______ ______

4. Count from 35 to 50 by 5s.

   ______ ______ ______ ______

5. Count from 60 to 75 by 5s.

   ______ ______ ______ ______

6. Count from 80 to 95 by 5s.

   ______ ______ ______ ______

Lesson 4

Name ____________

Counting by 1s, 5s, and 10s

You can count by 1s, 5s, or 10s.

Example

Count by 1s.	14, 15, 16, 17, 18, 19, 20, 21, 22, 23, 24
Count by 5s.	0, 5, 10, 15, 20, 25, 30, 35, 40, 45, 50, 55, 60, 65, 70, 75, 80, 85, 90, 95, 100, 105, 110, 115, 120
Count by 10s.	0, 10, 20, 30, 40, 50, 60, 70, 80, 90, 100, 110, 120

Count

Count out loud. Write the missing numbers.

1. Count by 1s.

   38 39 40 41 42

2. Count by 5s.

   85 90 ____ ____ ____

3. Count by 1s.

   106 107 ____ ____ ____

4. Count by 5s.

   25 30 ____ ____ ____

5. Count by 10s.

   40 50 ____ ____ ____

6. Count by 10s.

   80 90 ____ ____ ____

Name ______________________

Writing Numbers to 120 by 5s and 10s

You can count by 5s and 10s. You can write the numbers that you count.

Example

You can write numbers by 5s.

0 5 10 15 20 25 30

You can write numbers by 10s.

0 10 20 30 40

Count

Count by 5s. Write the missing numbers.

0	5	10	15		25
	35				55
	65		75		
90		100		110	115
120					

Count by 10s. Write the missing numbers.

0	10	20			50
	70		90		110
120					

Chapter 6 Test

Name ______________________________

Count on. Write the missing numbers.

Count by 1s. Write the missing numbers.

1. 103 104 ____ ____ ____
5. 56 57 ____ ____ ____
2. 106 107 ____ ____ ____
6. 67 68 ____ ____ ____
3. 114 115 ____ ____ ____
7. 97 98 ____ ____ ____
4. 41 42 ____ ____ ____
8. 75 76 ____ ____ ____

Count by 10s. Write the missing numbers.

9. 0 10 ____ ____ ____
13. 70 80 ____ ____ ____
10. 80 90 ____ ____ ____
14. 20 30 ____ ____ ____
11. 50 60 ____ ____ ____
15. 30 40 ____ ____ ____
12. 40 50 ____ ____ ____
16. 60 70 ____ ____ ____

Name ______________________

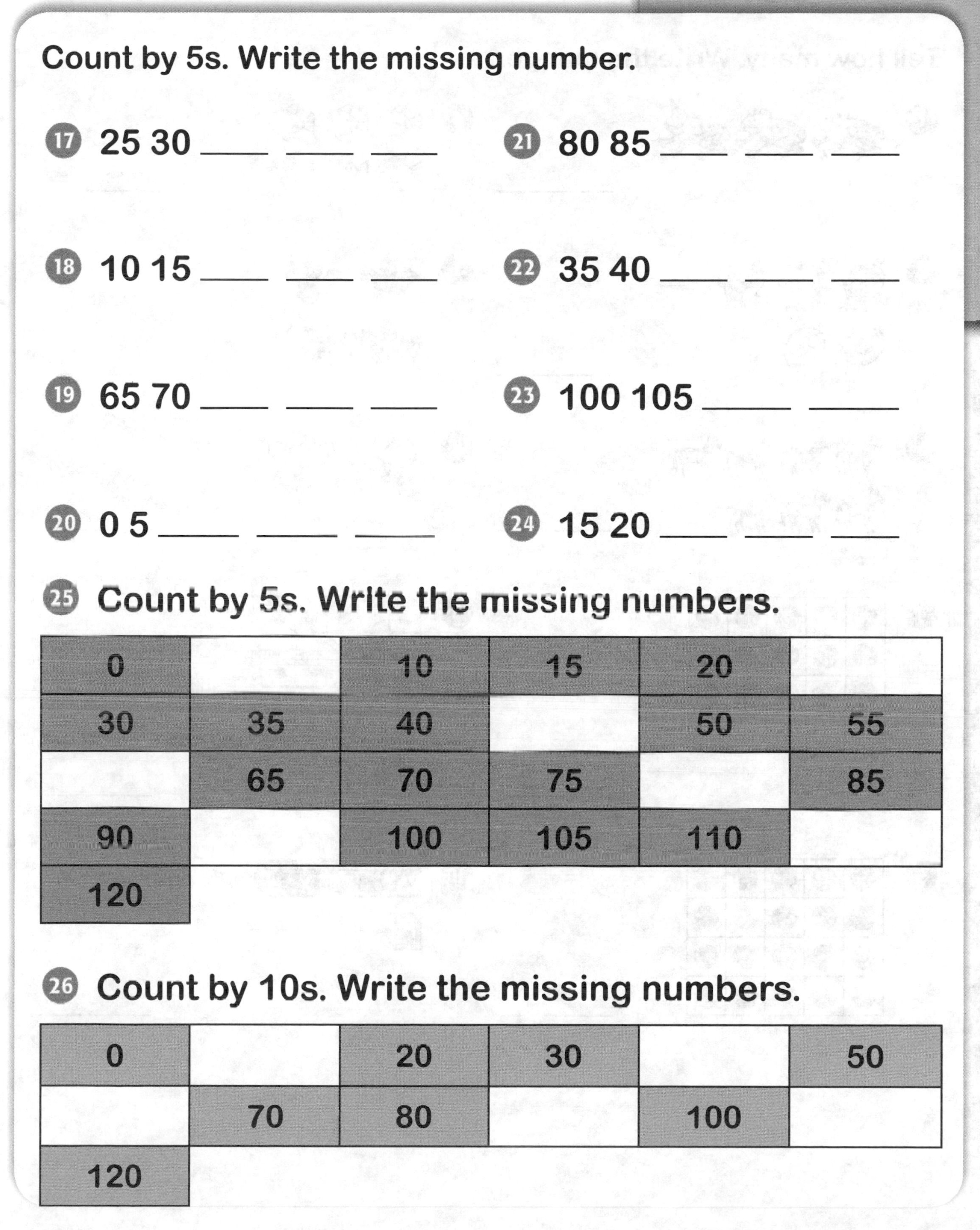

Count by 5s. Write the missing number.

17. 25 30 ____ ____ ____

18. 10 15 ____ ____ ____

19. 65 70 ____ ____ ____

20. 0 5 ____ ____ ____

21. 80 85 ____ ____ ____

22. 35 40 ____ ____ ____

23. 100 105 ____ ____

24. 15 20 ____ ____ ____

25. Count by 5s. Write the missing numbers.

0		10	15	20	
30	35	40		50	55
	65	70	75		85
90		100	105	110	
120					

26. Count by 10s. Write the missing numbers.

0		20	30		50
	70	80		100	
120					

Name ____________________

Tell how many. Write the number.

1  ________

2  ________

3  ________

4 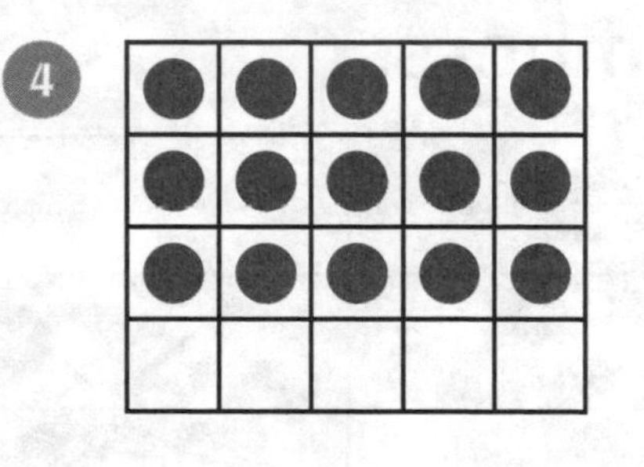 ________

5 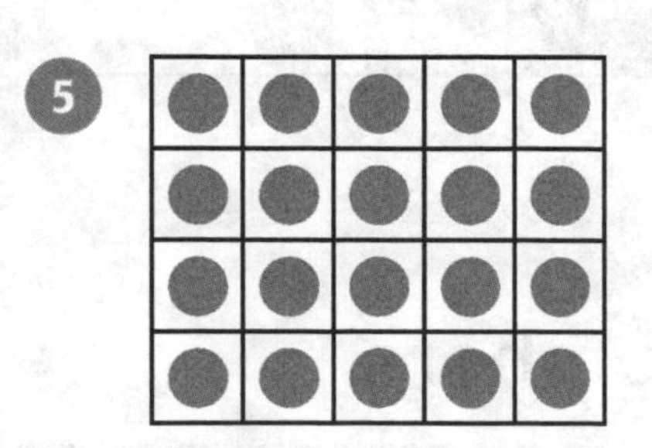 ________

6  ________

7  ________

8  ________

9 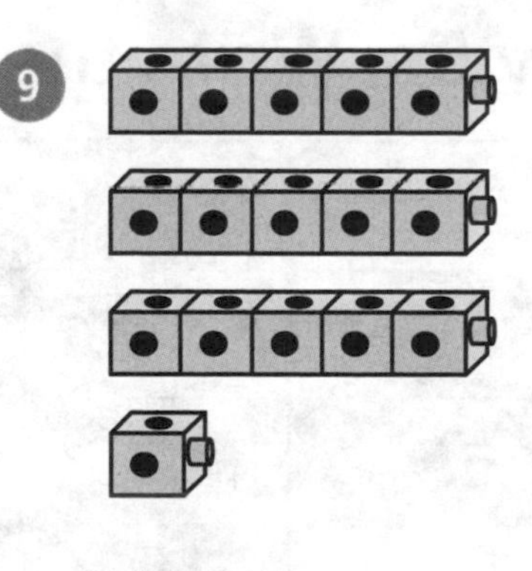 ________

10 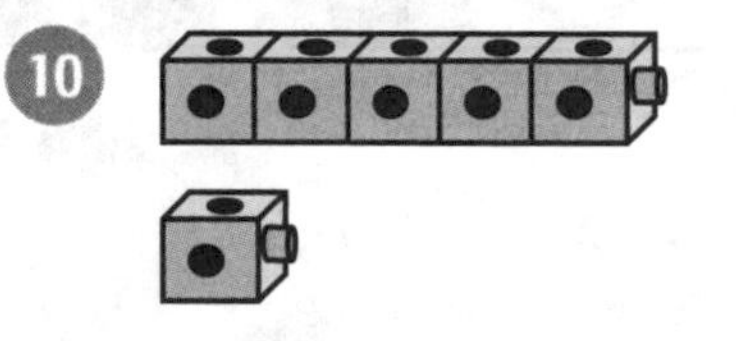 ________

Name ____________________

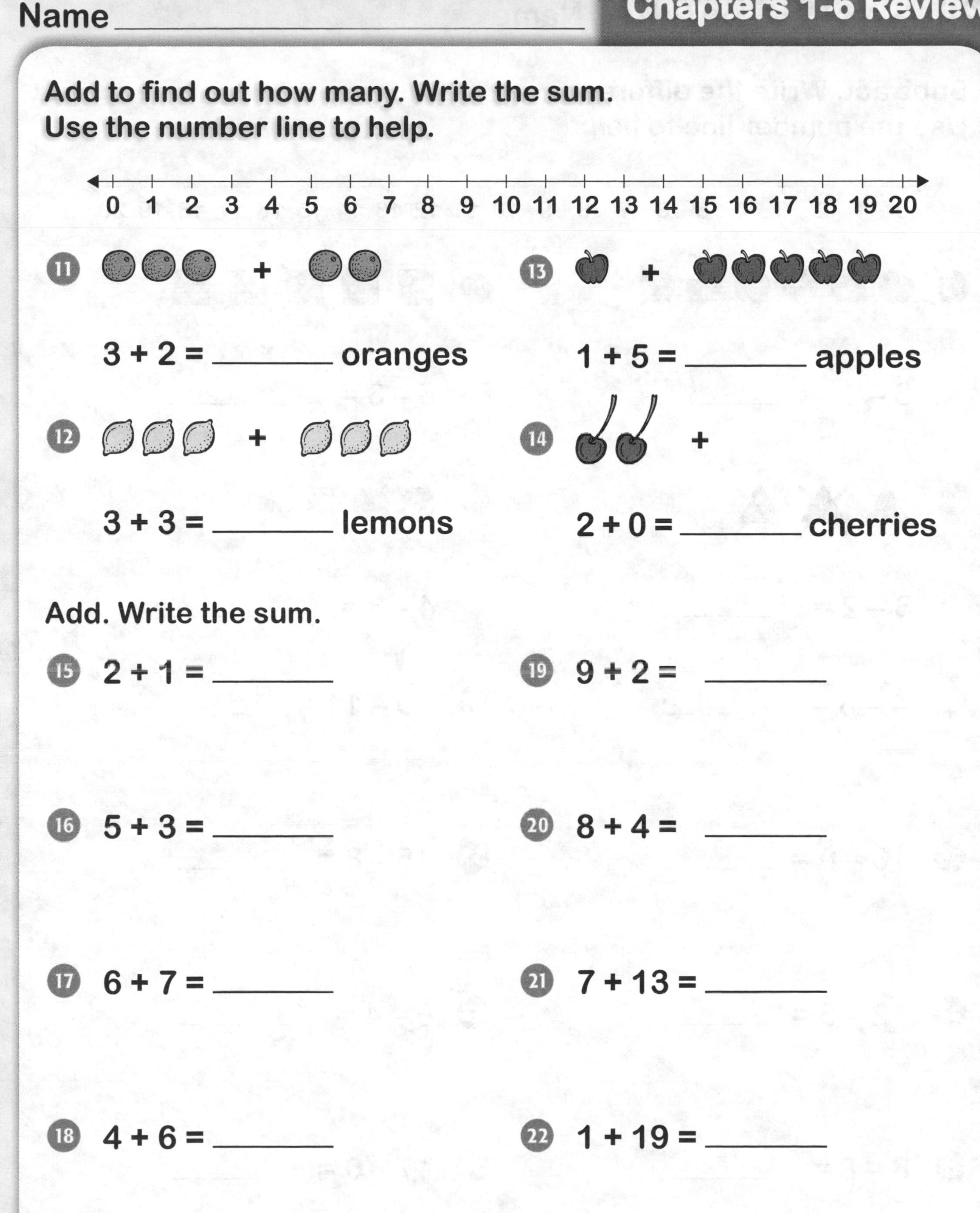

Add to find out how many. Write the sum.
Use the number line to help.

0 1 2 3 4 5 6 7 8 9 10 11 12 13 14 15 16 17 18 19 20

11. 3 + 2 = ______ oranges

13. 1 + 5 = ______ apples

12. 3 + 3 = ______ lemons

14. 2 + 0 = ______ cherries

Add. Write the sum.

15. 2 + 1 = ______

19. 9 + 2 = ______

16. 5 + 3 = ______

20. 8 + 4 = ______

17. 6 + 7 = ______

21. 7 + 13 = ______

18. 4 + 6 = ______

22. 1 + 19 = ______

Name ______________________

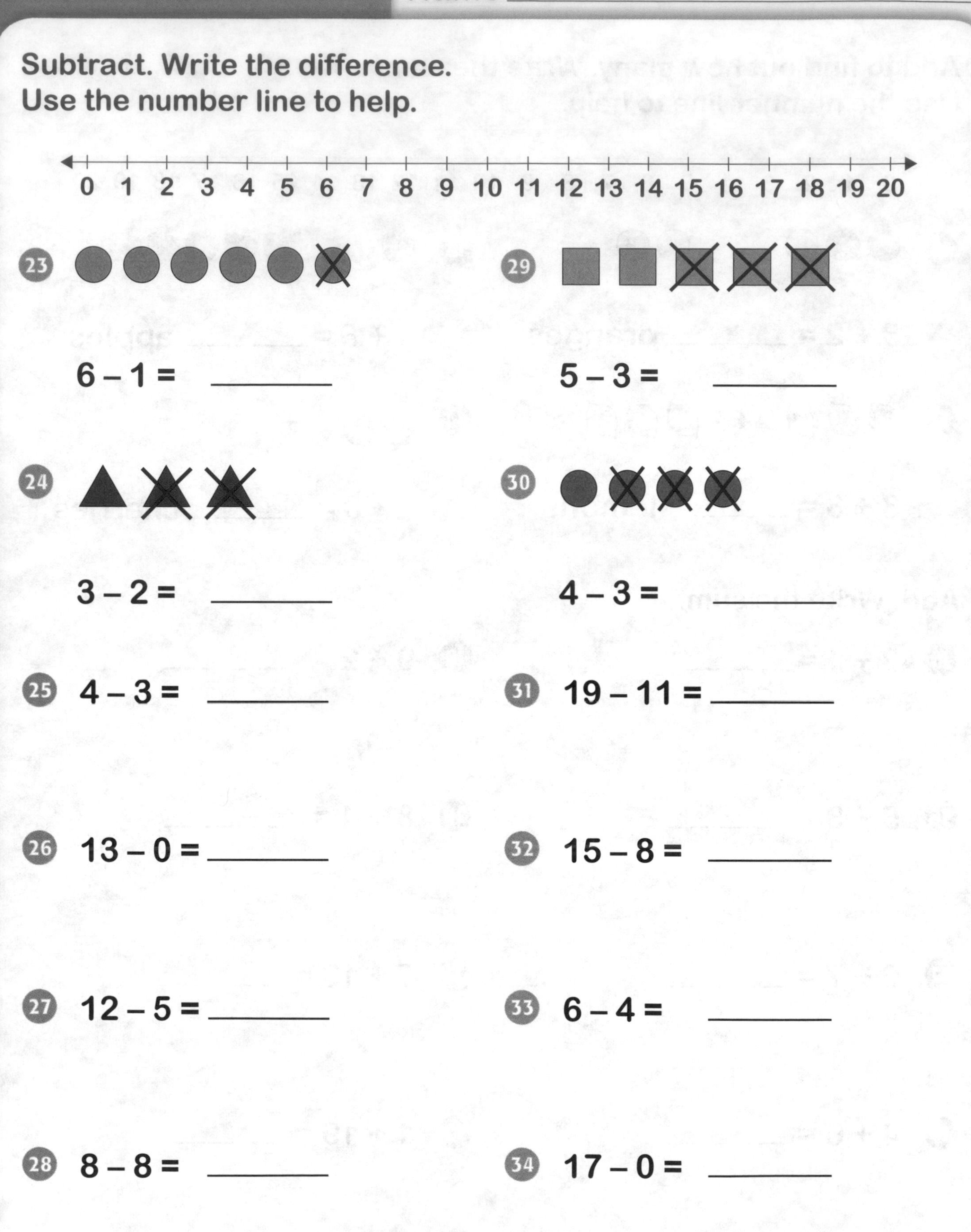

Subtract. Write the difference.
Use the number line to help.

0 1 2 3 4 5 6 7 8 9 10 11 12 13 14 15 16 17 18 19 20

23. $6 - 1 =$ ________

29. $5 - 3 =$ ________

24. $3 - 2 =$ ________

30. $4 - 3 =$ ________

25. $4 - 3 =$ ________

31. $19 - 11 =$ ________

26. $13 - 0 =$ ________

32. $15 - 8 =$ ________

27. $12 - 5 =$ ________

33. $6 - 4 =$ ________

28. $8 - 8 =$ ________

34. $17 - 0 =$ ________

Name ____________________

Count by 1s, 5s, or 10s. Write the missing numbers.

35. 5 6 7 ____ ____ ____ ____ 12

36. 65 70 ____ ____ 85 ____ ____

37. 10 20 ____ 40 ____ ____ ____

38. 25 30 ____ ____ ____ ____ 55

Solve. Write the sum or difference.

39. Elsa has 4 fish.
Nate has 5 fish.
How many fish in all?

4 + 5 = ____ fish

40. There are 8 cars.
2 cars leave.
How many cars are left?

8 – 2 = ____ cars

41. There are 19 birds.
3 birds fly away.
How many birds are left?

19 – 3 = ____ birds

42. There are 7 red blocks and 7 green blocks.
How many blocks in all?

7 + 7 = ____ blocks

Name ____________________

Adding in Different Orders

You can add numbers in different orders. The sum will be the same.

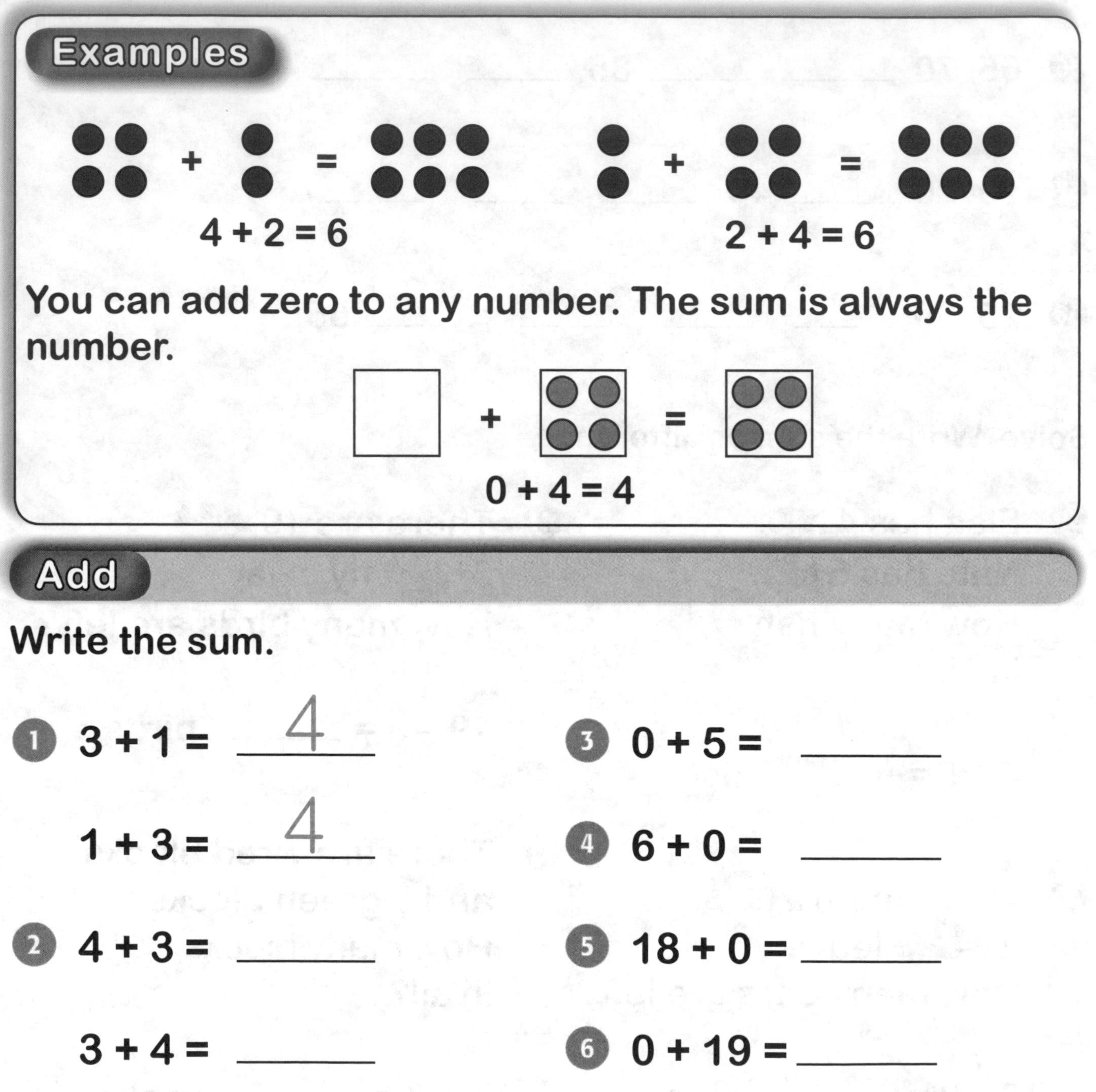

Examples

4 + 2 = 6

2 + 4 = 6

You can add zero to any number. The sum is always the number.

0 + 4 = 4

Add

Write the sum.

1. 3 + 1 = 4

   1 + 3 = 4

2. 4 + 3 = ______

   3 + 4 = ______

3. 0 + 5 = ______

4. 6 + 0 = ______

5. 18 + 0 = ______

6. 0 + 19 = ______

Name ____________________

Adding Three Numbers

You can add three numbers. You can group the numbers in any way. The sum will be the same.

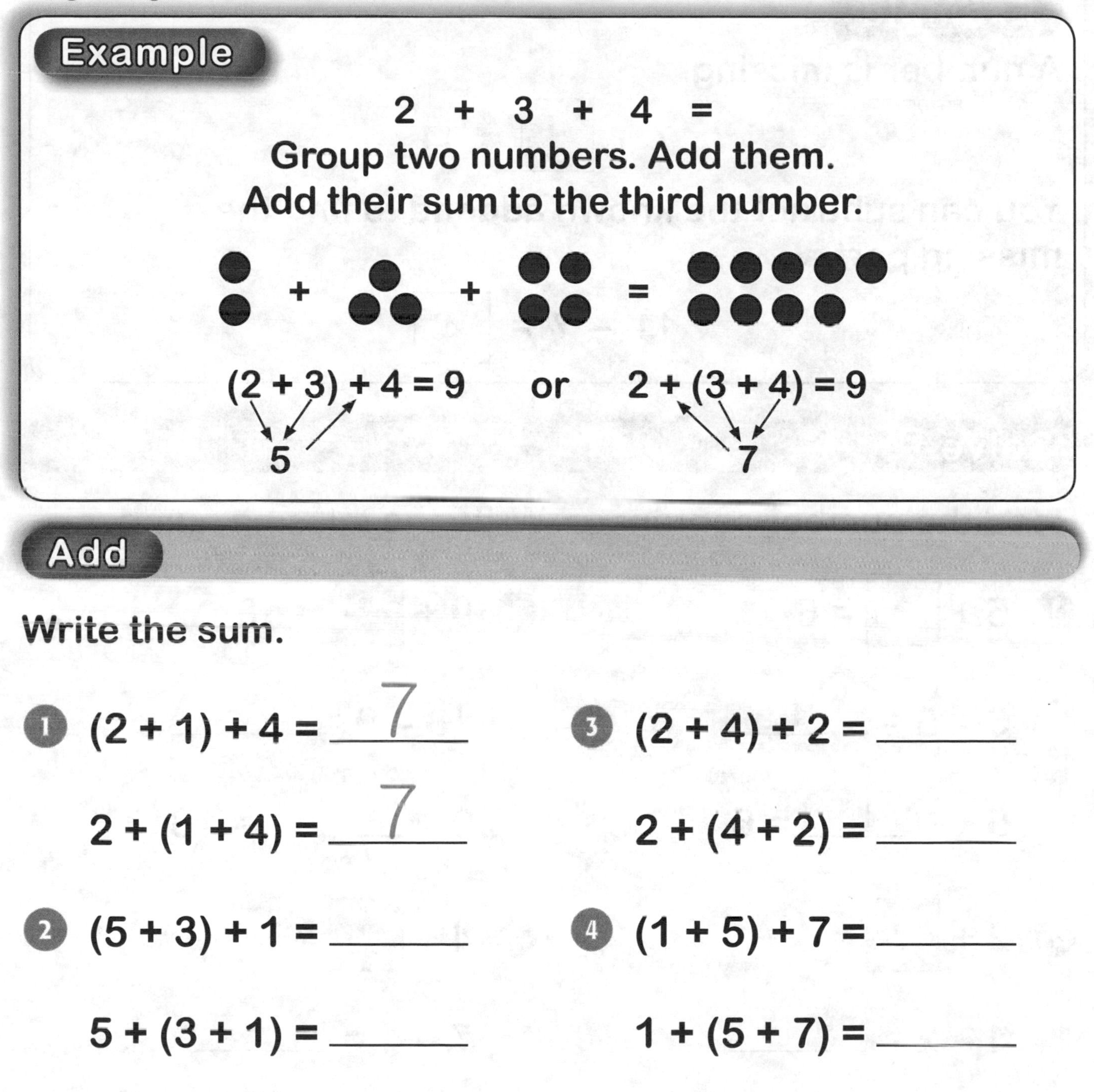

Example

2 + 3 + 4 =

Group two numbers. Add them.
Add their sum to the third number.

(2 + 3) + 4 = 9 or 2 + (3 + 4) = 9

5 7

Add

Write the sum.

1. (2 + 1) + 4 = 7

   2 + (1 + 4) = 7

2. (5 + 3) + 1 = ______

   5 + (3 + 1) = ______

3. (2 + 4) + 2 = ______

   2 + (4 + 2) = ______

4. (1 + 5) + 7 = ______

   1 + (5 + 7) = ______

Name ________________________

Subtract to Find Missing Numbers

You can find a missing addend.

Example

A number is missing.

7 + [?] = 11

You can subtract the known addend to find the missing part.

11 − 7 = [4]

Solve

Read the number sentences. Write the missing number.

1. 5 + [?] = 6

   6 − 5 = 1

   5 + 1 = 6

2. 2 + [?] = 4

   4 − 2 = ______

   2 + ______ = 4

3. 9 + [?] = 18

   18 − 9 = ______

   9 + ______ = 18

4. 4 + [?] = 7

   7 − 4 = ______

   4 + ______ = 7

Name ____________________

Counting to Add and Subtract

You can count on to add.

You can count back to subtract.

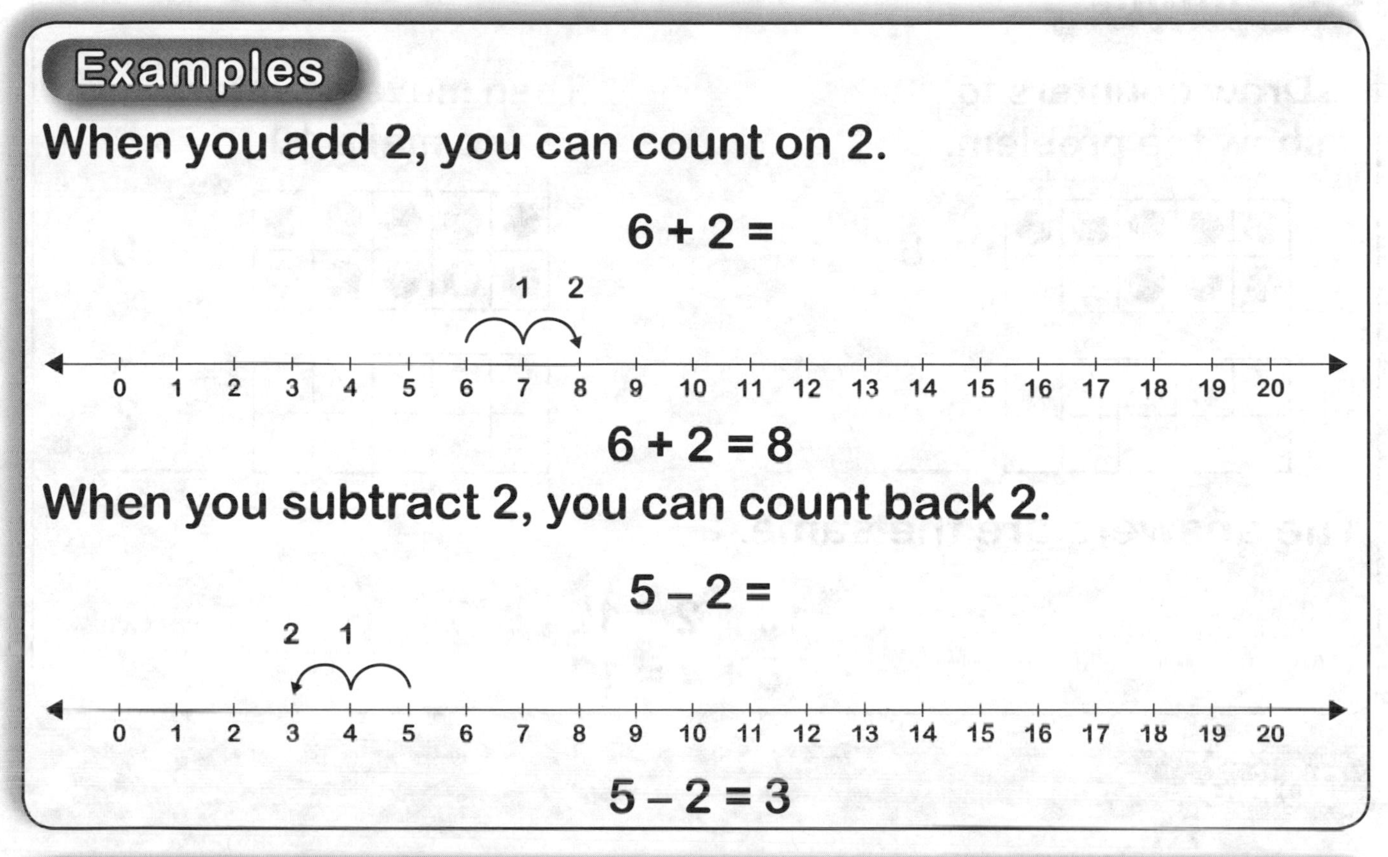

Examples

When you add 2, you can count on 2.

6 + 2 =

6 + 2 = 8

When you subtract 2, you can count back 2.

5 – 2 =

5 – 2 = 3

Add or Subtract

Count on to add. Count back to subtract. Write the sum or difference.

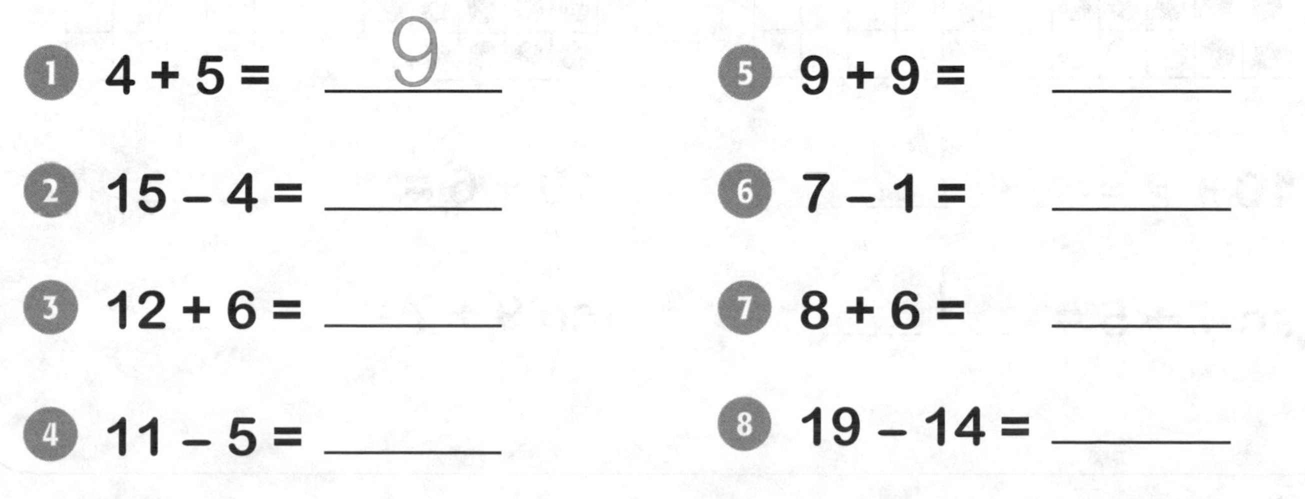

1. 4 + 5 = 9
2. 15 – 4 = ______
3. 12 + 6 = ______
4. 11 – 5 = ______
5. 9 + 9 = ______
6. 7 – 1 = ______
7. 8 + 6 = ______
8. 19 – 14 = ______

Name ______________________

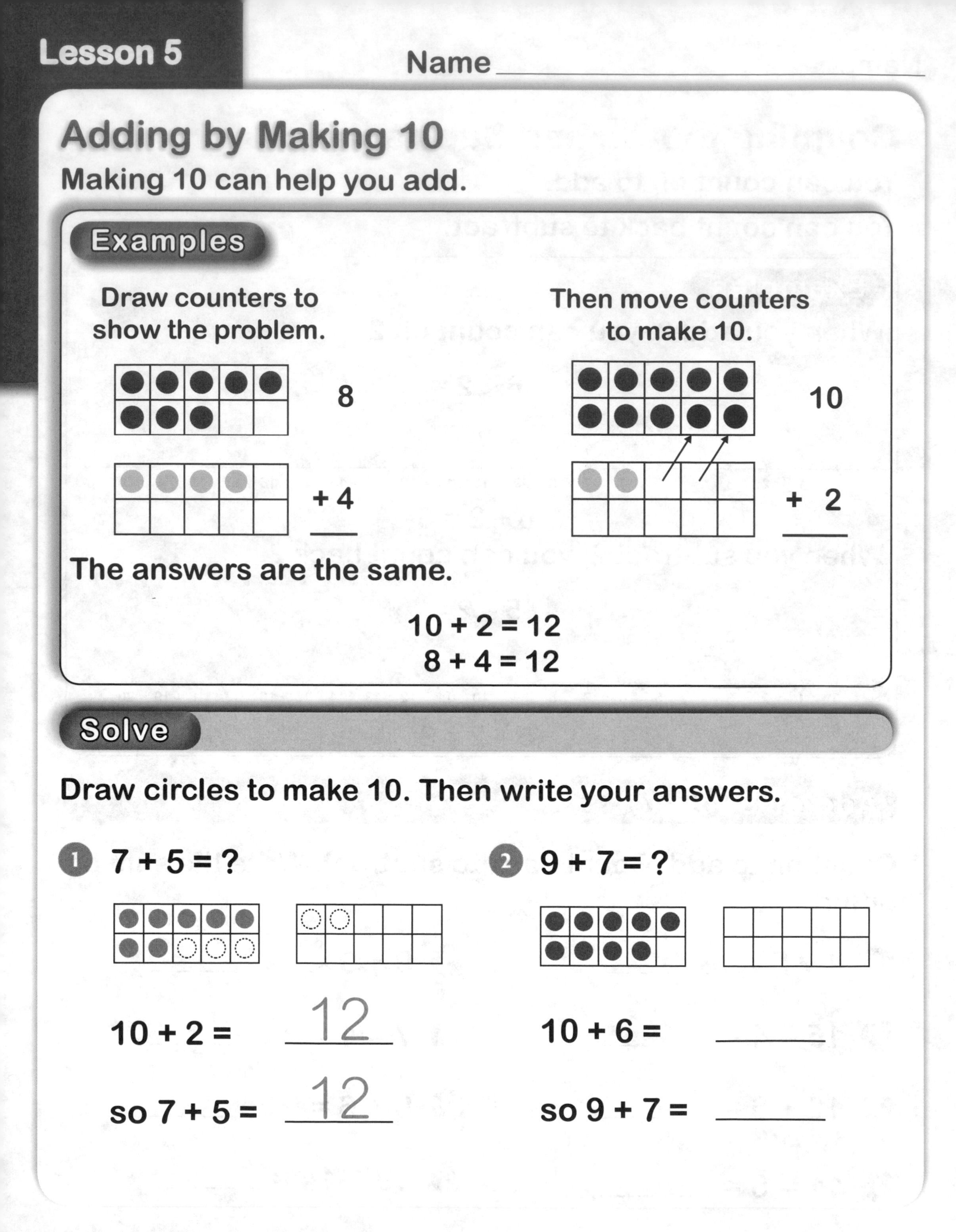

Adding by Making 10

Making 10 can help you add.

Examples

Draw counters to show the problem.

8
+ 4

Then move counters to make 10.

10
+ 2

The answers are the same.

10 + 2 = 12
8 + 4 = 12

Solve

Draw circles to make 10. Then write your answers.

1. 7 + 5 = ?

   10 + 2 = 12

   so 7 + 5 = 12

2. 9 + 7 = ?

   10 + 6 = ______

   so 9 + 7 = ______

Using Addition and Subtraction

When you know an addition fact, you also know a subtraction fact.

Example

This shows both addition and subtraction.

11 whole	
4 part	7 part

$4 + 7 = 11$
part part whole

$7 + 4 = 11$
part part whole

$11 - 7 = 4$
whole part part

$11 - 4 = 7$
whole part part

You add the parts to get the whole.

You subtract one part from the whole to get the other part.

Solve

Look at the chart. Write an addition sentence. Write a subtraction sentence.

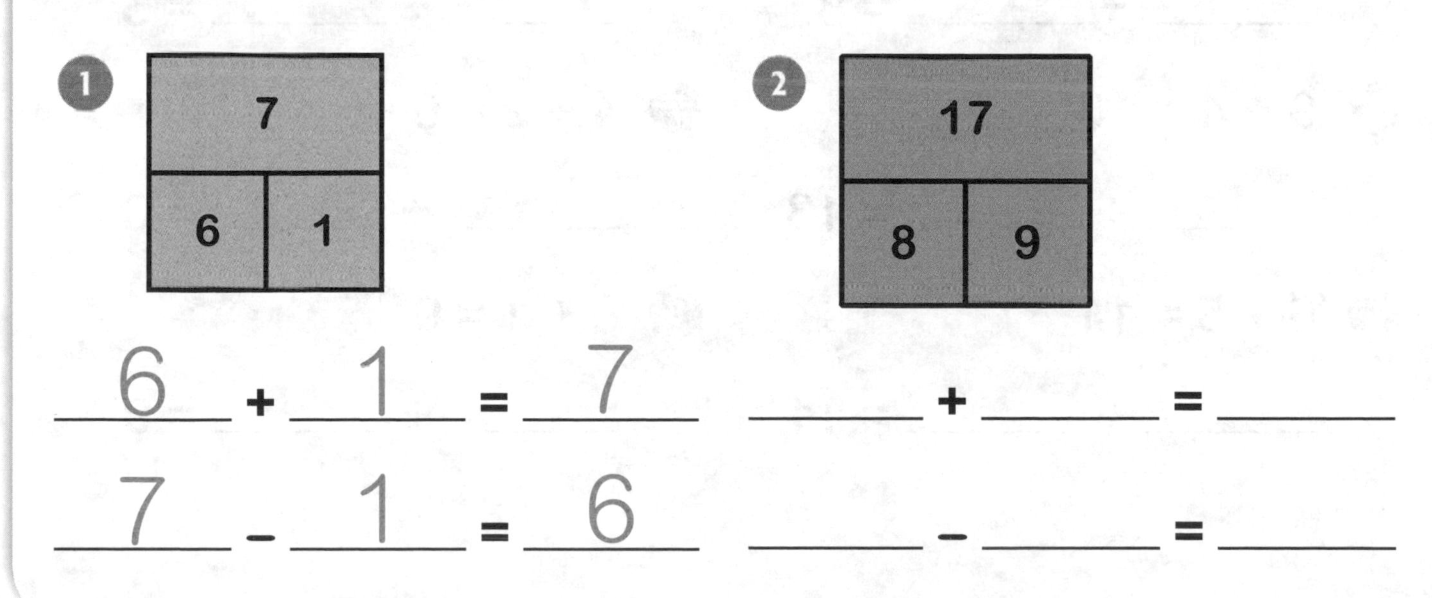

Name ____________________

Creating New Sums

Different numbers add up to the same sum.

This can help you add.

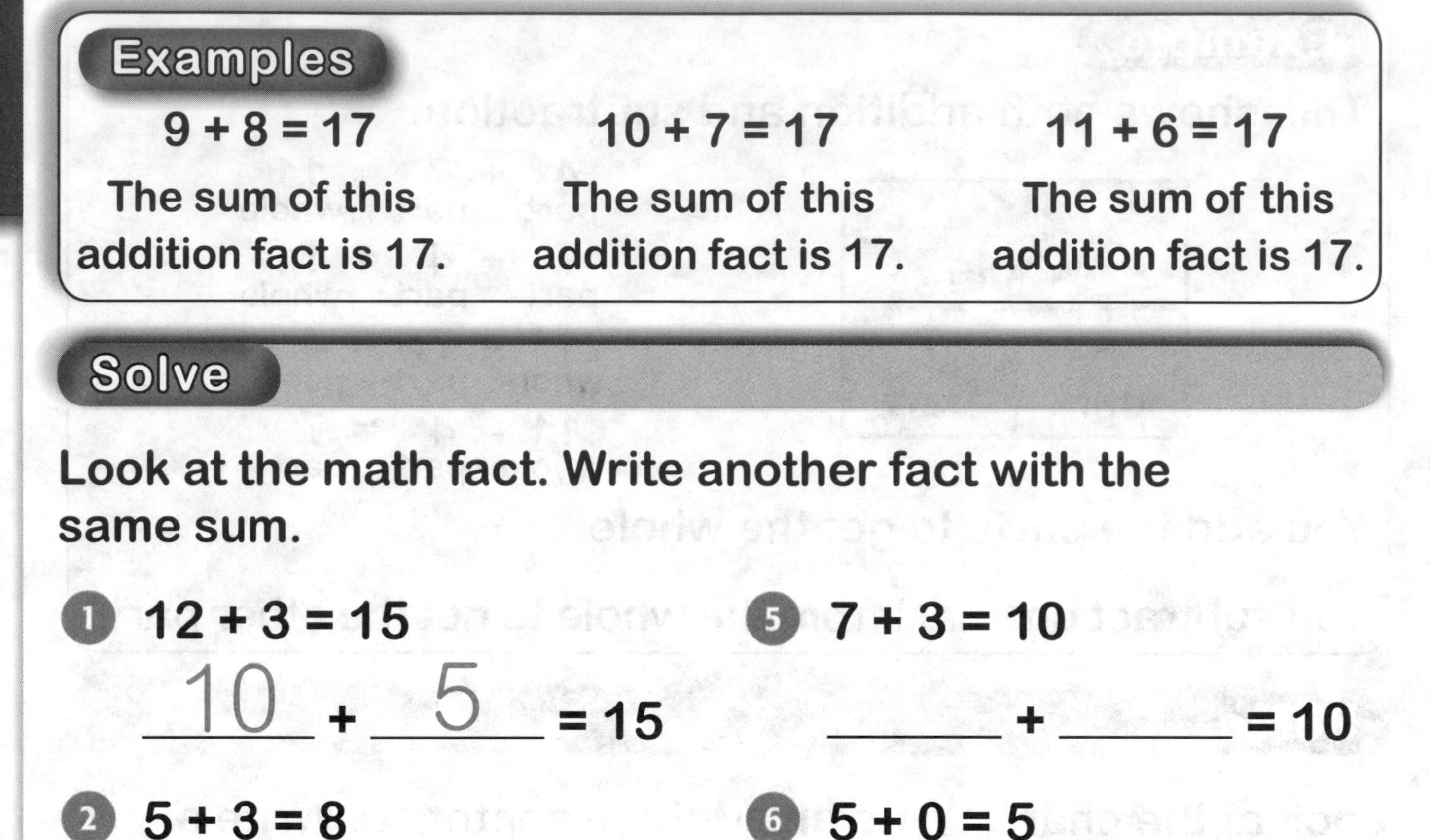

Examples

$9 + 8 = 17$	$10 + 7 = 17$	$11 + 6 = 17$
The sum of this addition fact is 17.	The sum of this addition fact is 17.	The sum of this addition fact is 17.

Solve

Look at the math fact. Write another fact with the same sum.

1. $12 + 3 = 15$

   10 + 5 = 15

2. $5 + 3 = 8$

   ______ + ______ = 8

3. $9 + 4 = 13$

   ______ + ______ = 13

4. $6 + 5 = 11$

   ______ + ______ = 11

5. $7 + 3 = 10$

   ______ + ______ = 10

6. $5 + 0 = 5$

   ______ + ______ = 5

7. $4 + 2 = 6$

   ______ + ______ = 6

8. $8 + 1 = 9$

   ______ + ______ = 9

Name ____________________

Solving Word Problems

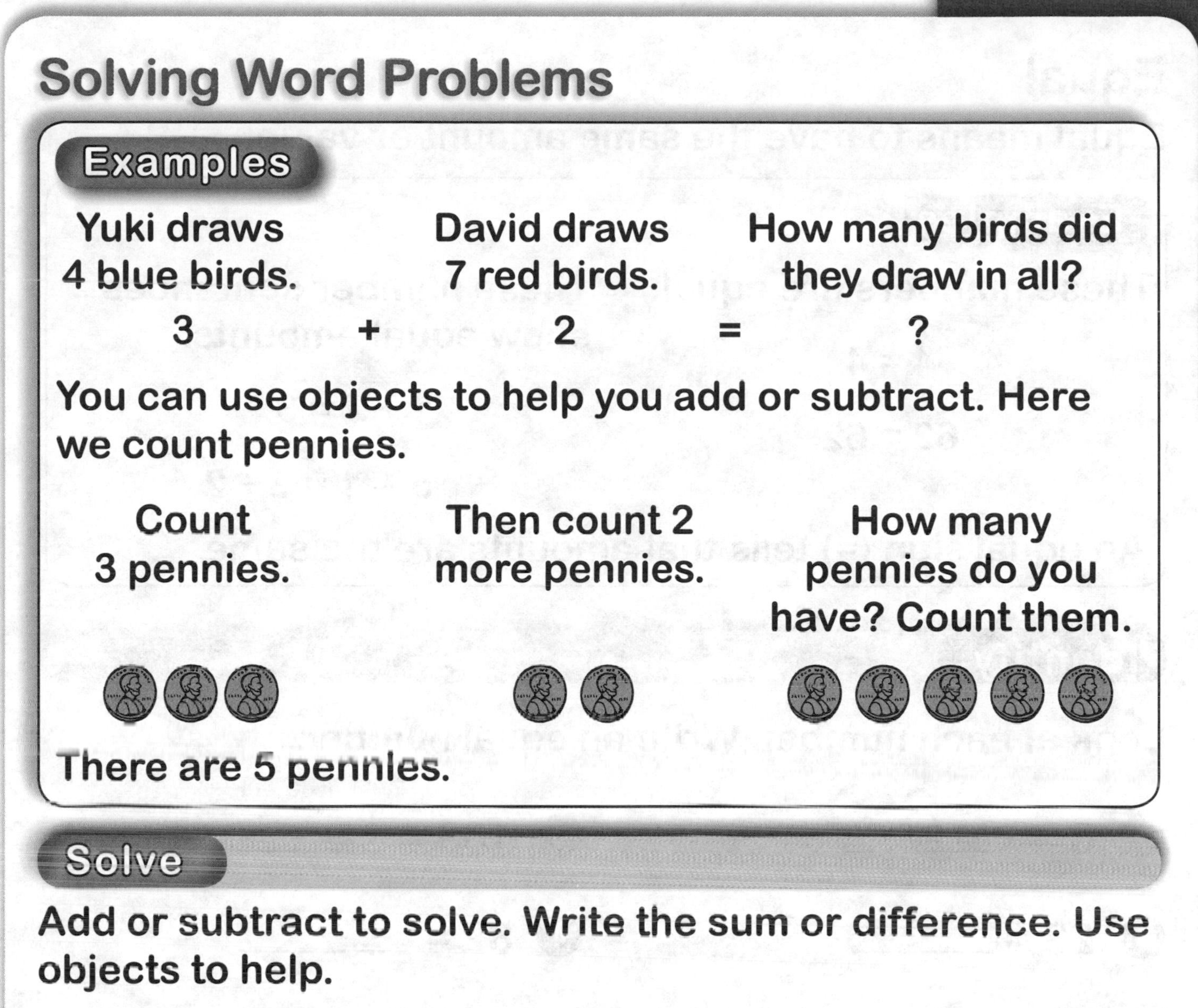

Examples

Yuki draws 4 blue birds.	David draws 7 red birds.	How many birds did they draw in all?
3 +	2 =	?

You can use objects to help you add or subtract. Here we count pennies.

Count 3 pennies.	Then count 2 more pennies.	How many pennies do you have? Count them.

There are 5 pennies.

Solve

Add or subtract to solve. Write the sum or difference. Use objects to help.

1. There are 14 bees.
   5 bees fly away.
   How many bees are left?

   14 – 5 = __9__

2. Ella draws 4 stars.
   She draws 2 more stars.
   How many stars does Ella draw?

   4 + 2 = ______

Name ____________________

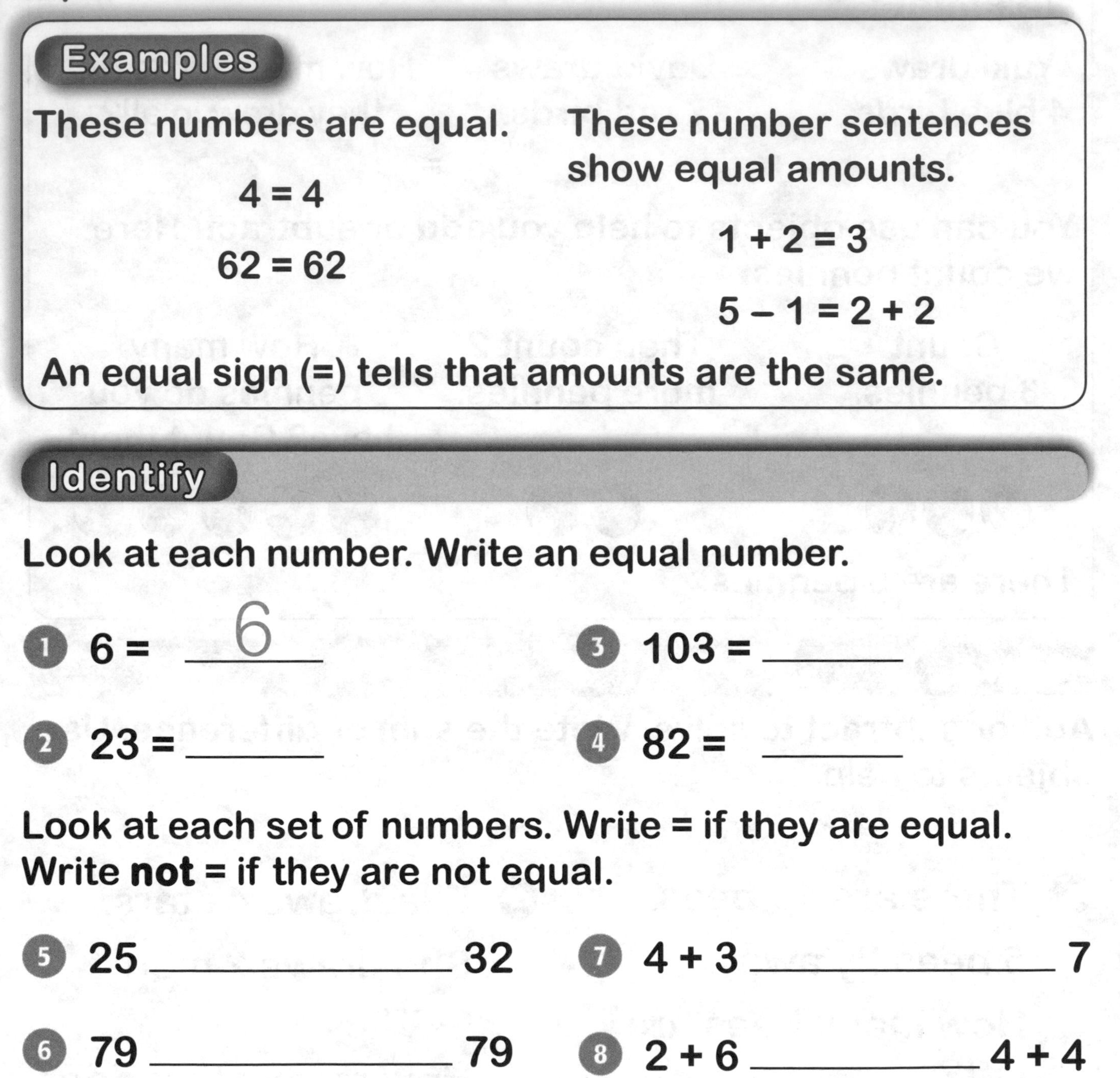

Equal

Equal means to have the same amount or value.

Examples

These numbers are equal.

4 = 4

62 = 62

These number sentences show equal amounts.

1 + 2 = 3

5 – 1 = 2 + 2

An equal sign (=) tells that amounts are the same.

Identify

Look at each number. Write an equal number.

1. 6 = 6
2. 23 = ______
3. 103 = ______
4. 82 = ______

Look at each set of numbers. Write = if they are equal. Write **not** = if they are not equal.

5. 25 ____________ 32
6. 79 ____________ 79
7. 4 + 3 ____________ 7
8. 2 + 6 ____________ 4 + 4

Name ____________________

Read the number sentences. Write the sums.

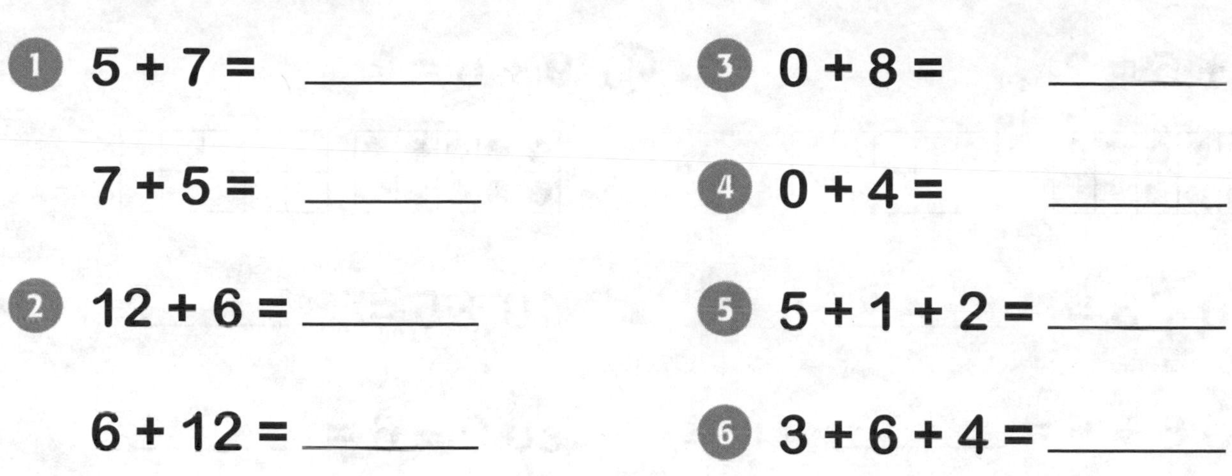

1. $5 + 7 =$ ______

   $7 + 5 =$ ______

2. $12 + 6 =$ ______

   $6 + 12 =$ ______

3. $0 + 8 =$ ______

4. $0 + 4 =$ ______

5. $5 + 1 + 2 =$ ______

6. $3 + 6 + 4 =$ ______

Read the problems. Write the missing number.

7. $5 +$ [?] $= 9$

   $9 - 5 =$ ______

   $5 +$ ______ $= 9$

8. $5 +$ [?] $= 18$

   $18 - 5 =$ ______

   $5 +$ ______ $= 18$

Look at each addition sentence. Write another fact with the same sum.

9. $9 + 2 = 11$

   ______ + ______ $= 11$

10. $3 + 4 = 7$

    ______ + ______ $= 7$

Look at each set of numbers. Write = if they are equal. Write **not** = if they are not equal.

11. 17 ______ 17

12. 35 ______ $30 + 5$

13. $8 + 3$ ______ $12 + 5$

14. $9 - 3$ ______ $4 + 2$

Name ____________________

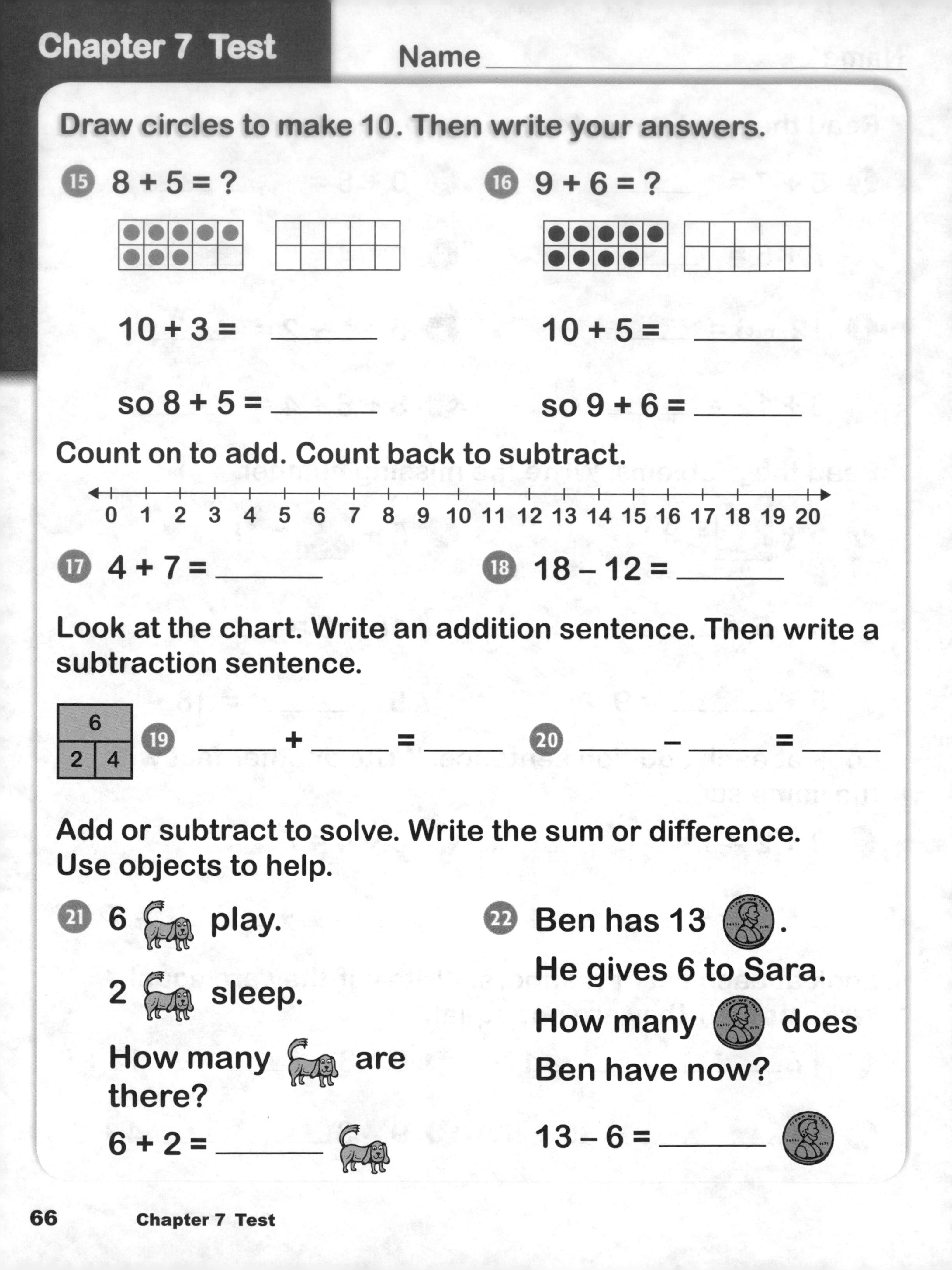

Draw circles to make 10. Then write your answers.

15 8 + 5 = ?

10 + 3 = ______

so 8 + 5 = ______

16 9 + 6 = ?

10 + 5 = ______

so 9 + 6 = ______

Count on to add. Count back to subtract.

17 4 + 7 = ______

18 18 – 12 = ______

Look at the chart. Write an addition sentence. Then write a subtraction sentence.

6	
2	4

19 ____ + ____ = ____

20 ____ – ____ = ____

Add or subtract to solve. Write the sum or difference. Use objects to help.

21 6 play.

2 sleep.

How many are there?

6 + 2 = ______

22 Ben has 13 .

He gives 6 to Sara.

How many does Ben have now?

13 – 6 = ______

Name ______________________

Thinking of 10

Examples

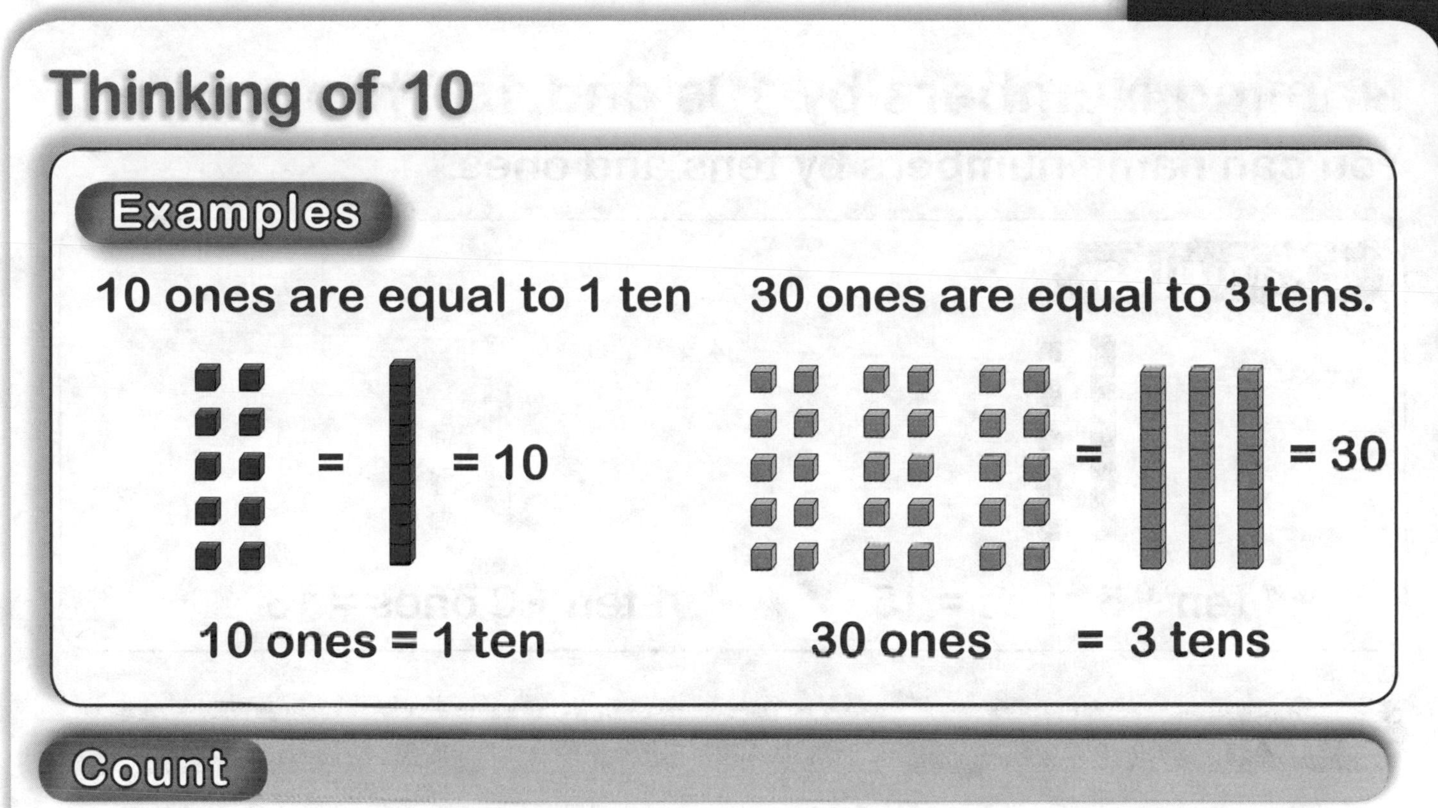

Count

Count the ones. Write how many ones. Write how many tens.

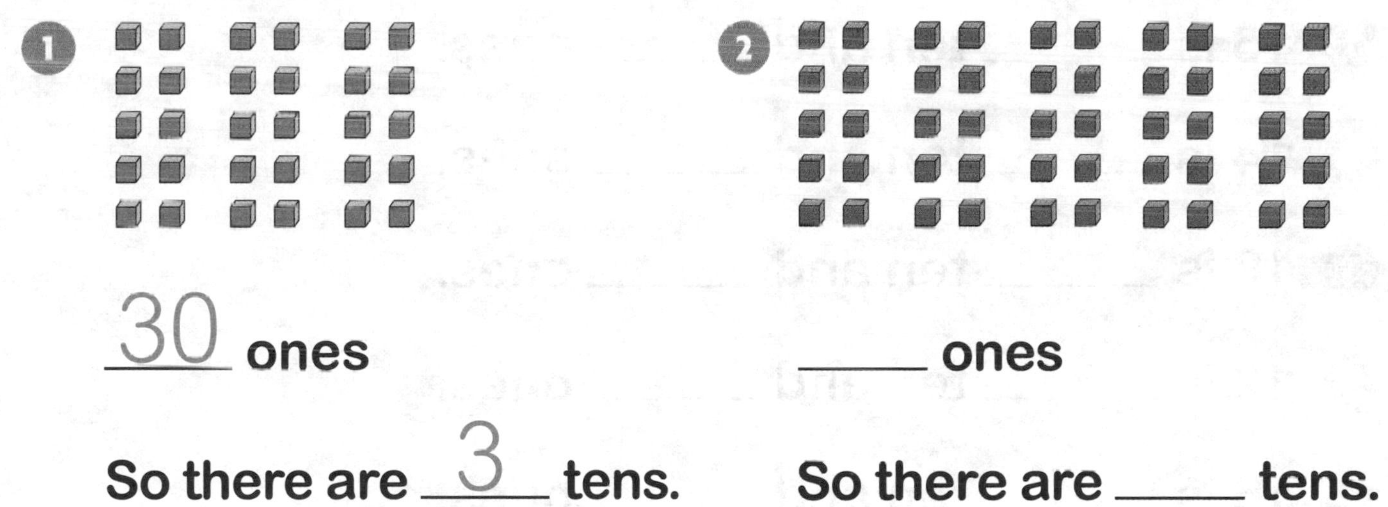

1. 30 ones

So there are 3 tens.

2. _____ ones

So there are _____ tens.

Name ______________________

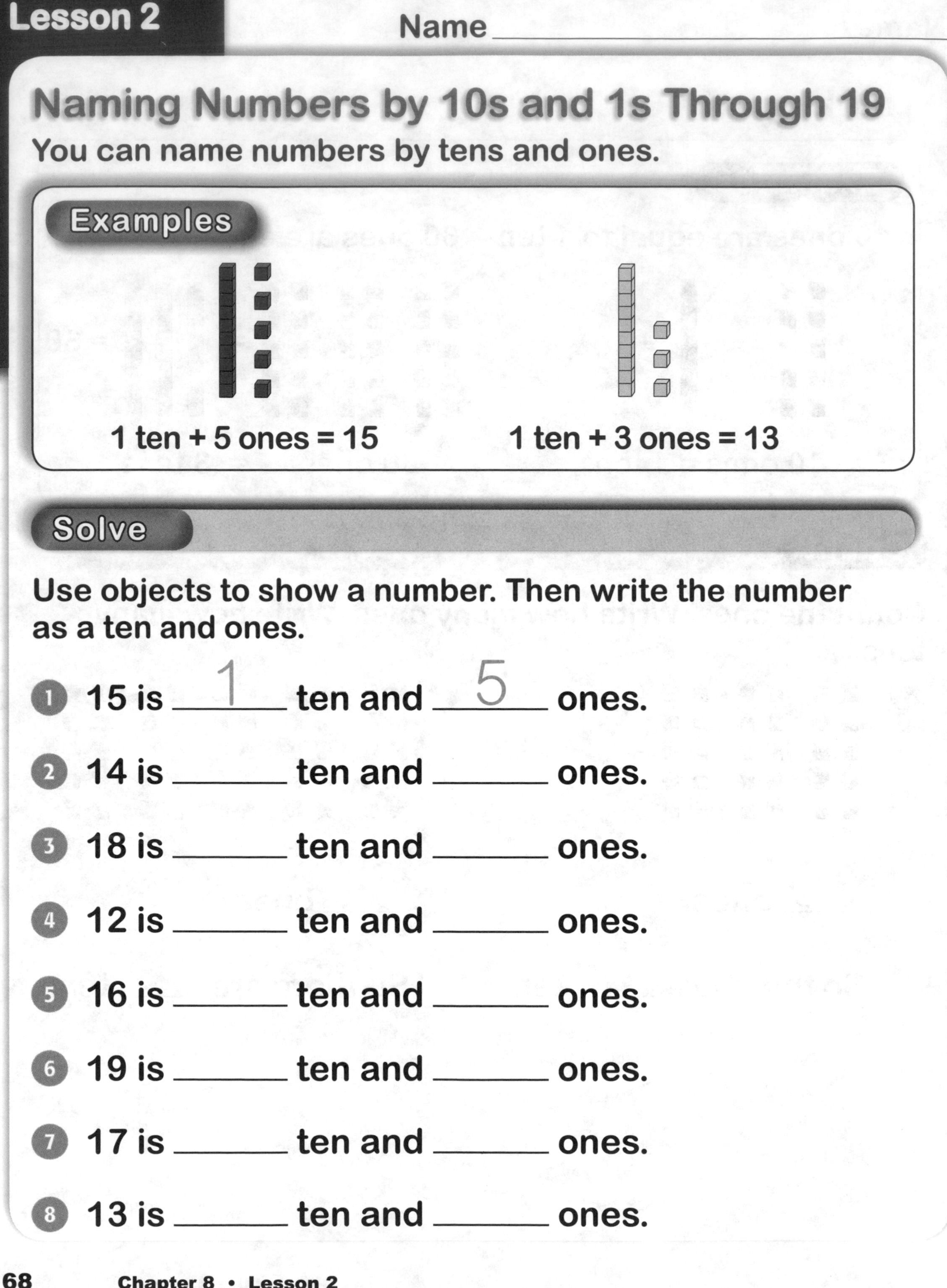

Naming Numbers by 10s and 1s Through 19

You can name numbers by tens and ones.

Examples

1 ten + 5 ones = 15

1 ten + 3 ones = 13

Solve

Use objects to show a number. Then write the number as a ten and ones.

1. 15 is __1__ ten and __5__ ones.
2. 14 is ______ ten and ______ ones.
3. 18 is ______ ten and ______ ones.
4. 12 is ______ ten and ______ ones.
5. 16 is ______ ten and ______ ones.
6. 19 is ______ ten and ______ ones.
7. 17 is ______ ten and ______ ones.
8. 13 is ______ ten and ______ ones.

Name ______________________

Naming Numbers by 10s Through 90

You can use tens to make numbers.

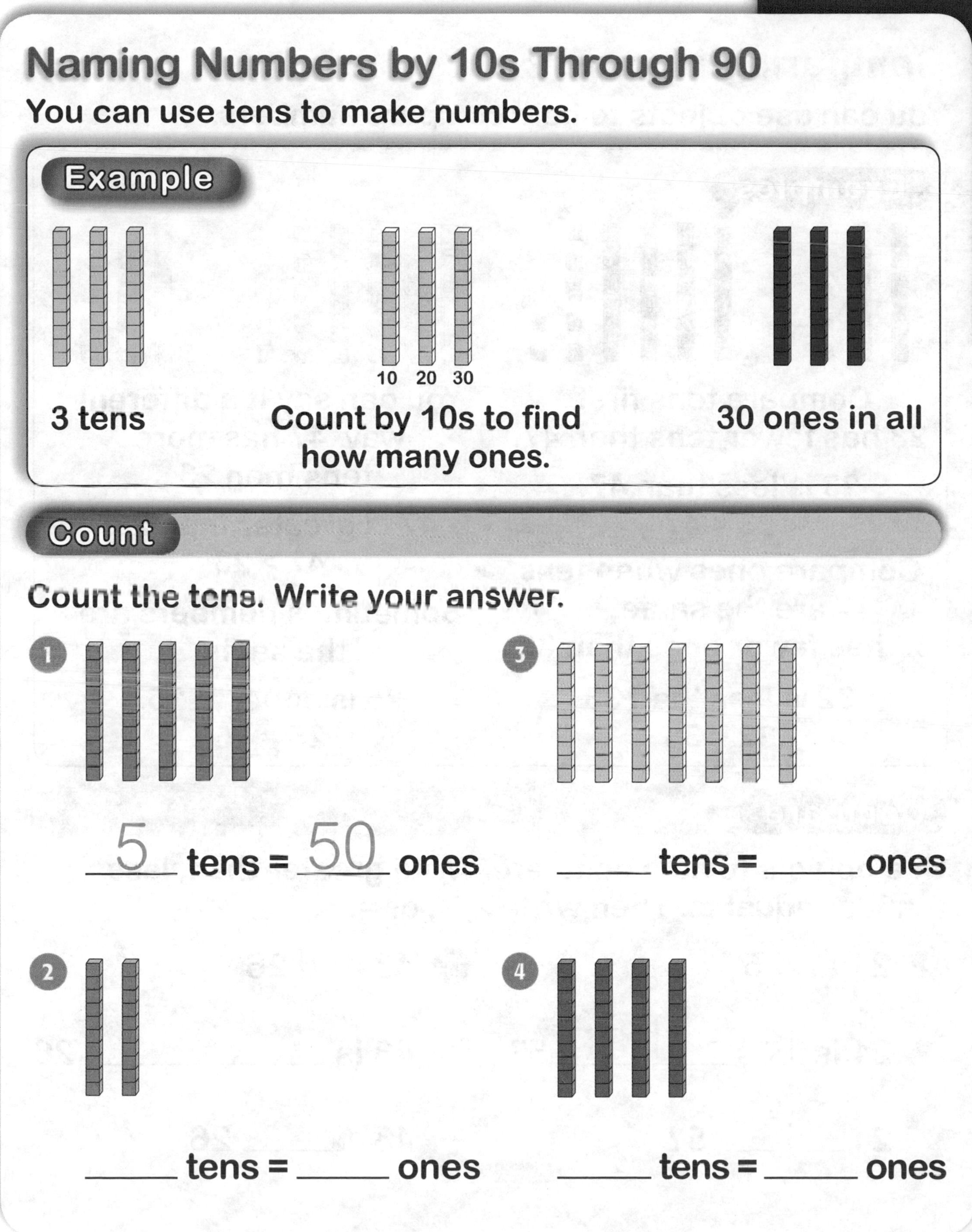

Count the tens. Write your answer.

1. 5 tens = 50 ones
2. ______ tens = ______ ones
3. ______ tens = ______ ones
4. ______ tens = ______ ones

Name ____________________

Comparing Numbers

You can use objects to help compare numbers.

Examples

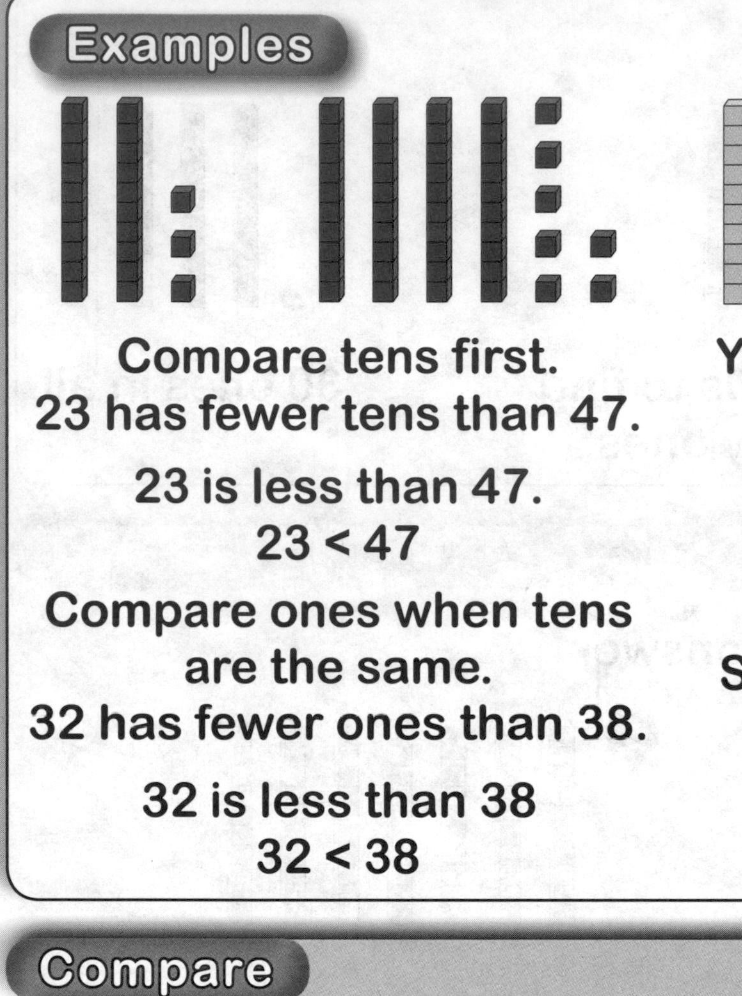

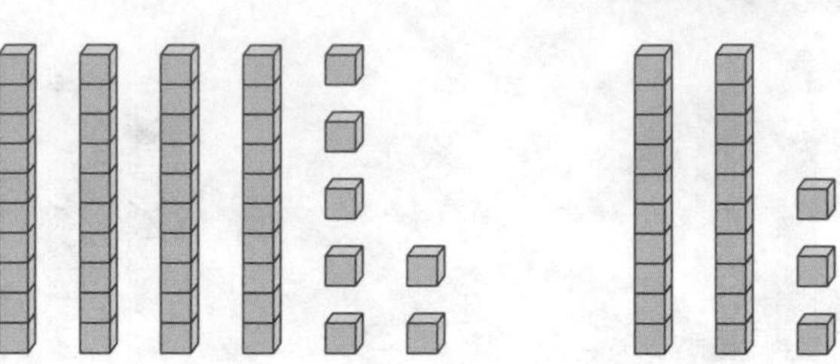

Compare tens first.
23 has fewer tens than 47.

23 is less than 47.
23 < 47

Compare ones when tens are the same.
32 has fewer ones than 38.

32 is less than 38
32 < 38

You can say it a different way. 47 has more tens than 23.

47 is greater than 23.
47 > 23

Sometimes numbers are the same.

25 is equal to 25.
25 = 25

Compare

Use objects to help compare. Write **greater than, less than**, or **equal to**. Then write >, <, or =.

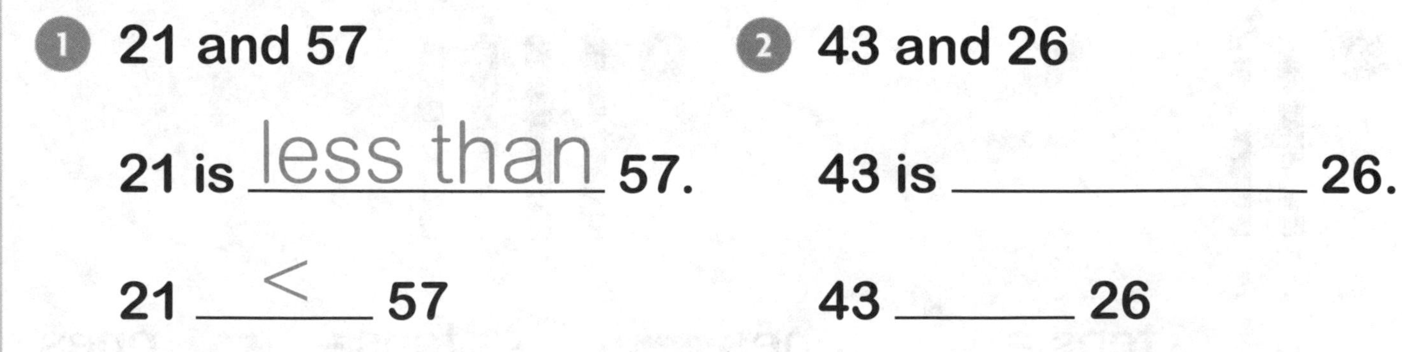

1. 21 and 57

   21 is less than 57.

   21 < 57

2. 43 and 26

   43 is ____________ 26.

   43 ________ 26

Name ____________________

Count the ones. Write how many ones. Write how many tens.

1. There are _____ ones.

So there are _____ tens.

2. There are _____ ones.

So there are _____ tens.

Use paper clips to make each number. Write the number as a ten and ones.

3. 13 is _____ ten and _____ ones.

4. 17 is _____ ten and _____ ones.

5. 11 is _____ ten and _____ one.

6. 19 is _____ ten and _____ ones.

Name ____________________

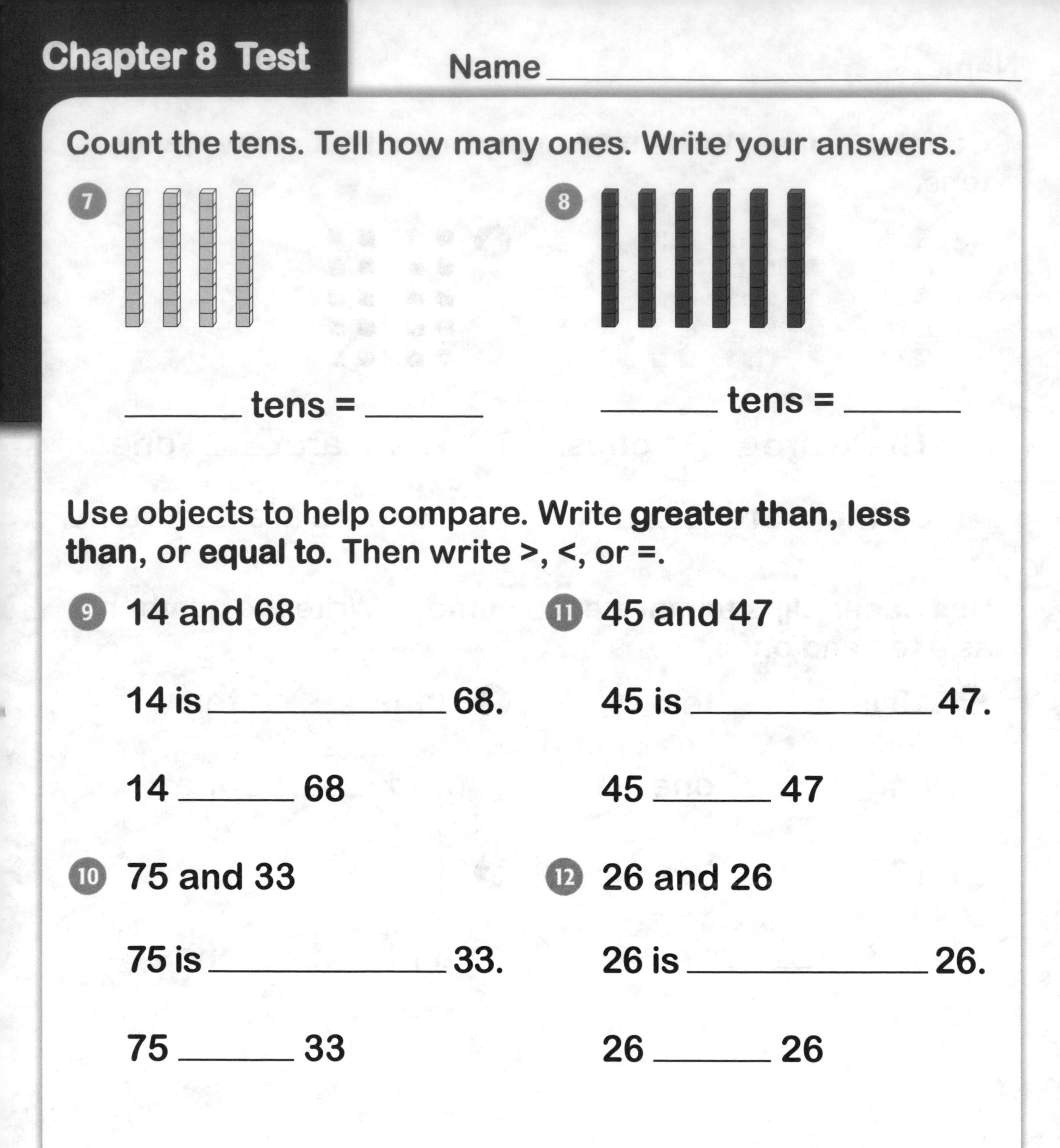

Count the tens. Tell how many ones. Write your answers.

7 ______ tens = ______

8 ______ tens = ______

Use objects to help compare. Write **greater than, less than,** or **equal to**. Then write >, <, or =.

9 14 and 68

14 is ____________ 68.

14 ______ 68

10 75 and 33

75 is ____________ 33.

75 ______ 33

11 45 and 47

45 is ____________ 47.

45 ______ 47

12 26 and 26

26 is ____________ 26.

26 ______ 26

Name ____________________

Adding Digits

Numbers have digits.

5 This number has one digit. It shows 5 ones.

23 This number has two digits. It shows 2 tens and 3 ones.

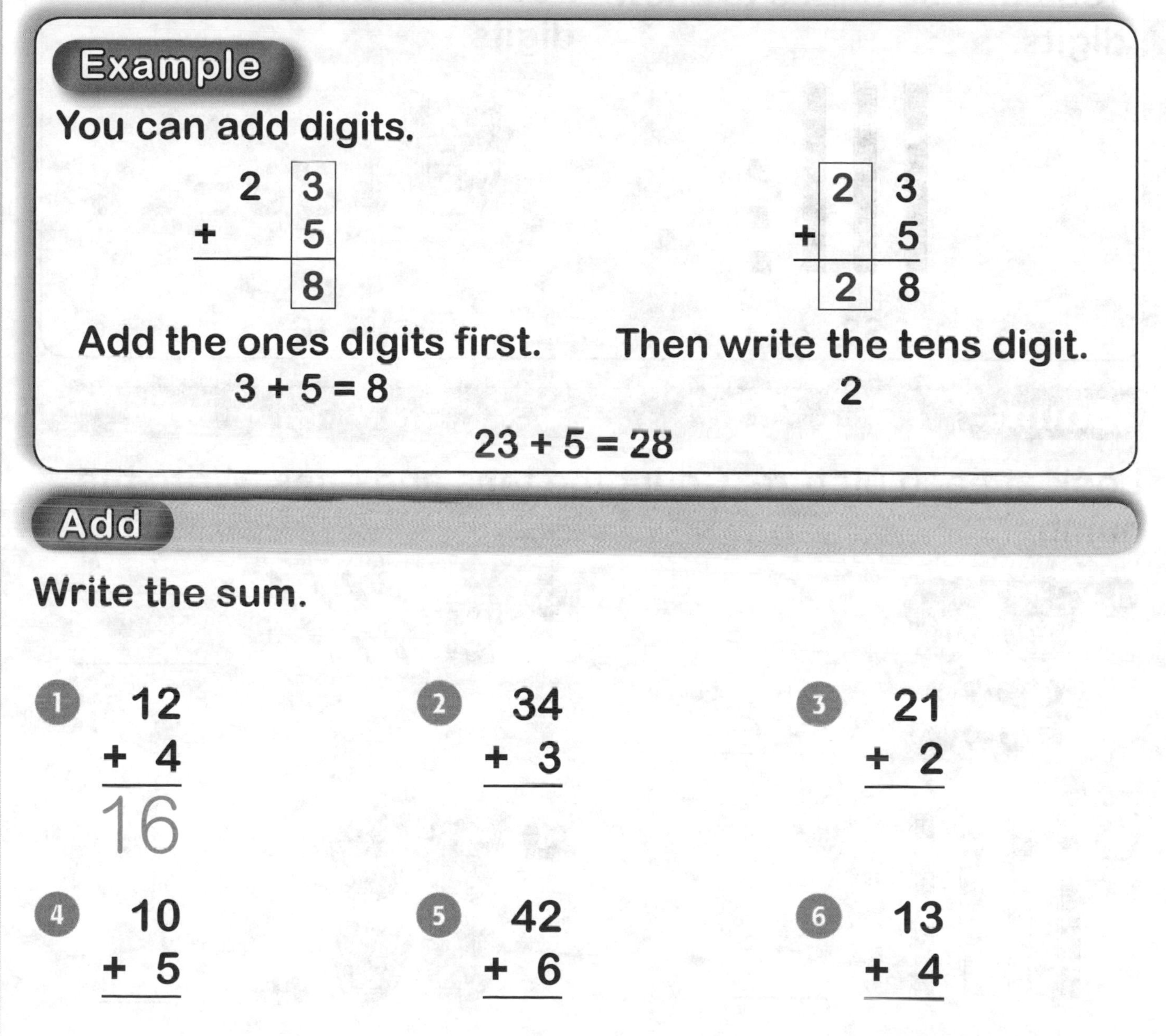

Example

You can add digits.

$$\begin{array}{r} 2\ \boxed{3} \\ +\ \ \boxed{5} \\ \hline \boxed{8} \end{array} \qquad \begin{array}{r} \boxed{2}\ 3 \\ +\ \ \ \ 5 \\ \hline \boxed{2}\ 8 \end{array}$$

Add the ones digits first.
$3 + 5 = 8$

Then write the tens digit.
2

$23 + 5 = 28$

Add

Write the sum.

1. $\begin{array}{r} 12 \\ +\ 4 \\ \hline 16 \end{array}$

2. $\begin{array}{r} 34 \\ +\ 3 \\ \hline \end{array}$

3. $\begin{array}{r} 21 \\ +\ 2 \\ \hline \end{array}$

4. $\begin{array}{r} 10 \\ +\ 5 \\ \hline \end{array}$

5. $\begin{array}{r} 42 \\ +\ 6 \\ \hline \end{array}$

6. $\begin{array}{r} 13 \\ +\ 4 \\ \hline \end{array}$

Lesson 2

Name ______________________

Show Tens and Ones

You can use objects and pictures to show numbers.

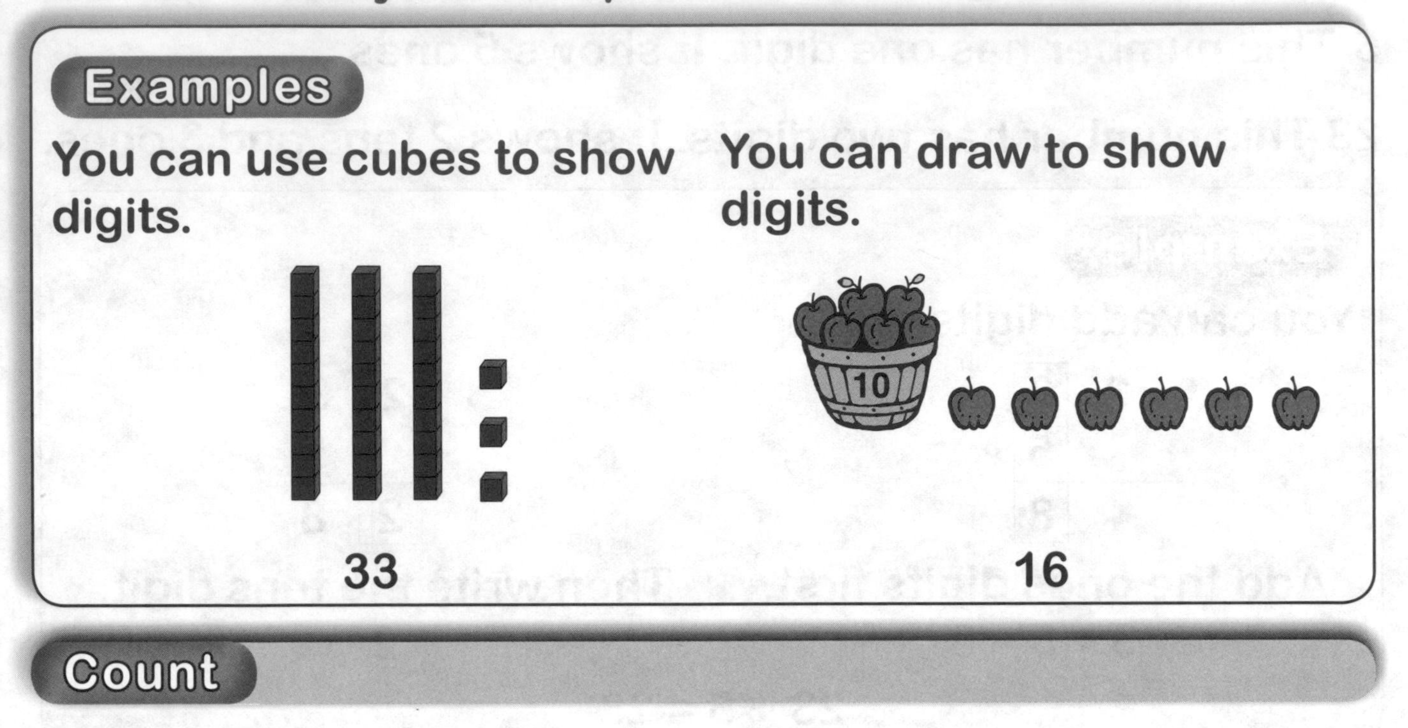

Count

Look at each picture. Count the tens and ones. Write the number.

Name ______________________________

Showing Tens and Ones on a Chart

You can show numbers on a Place Value Chart.

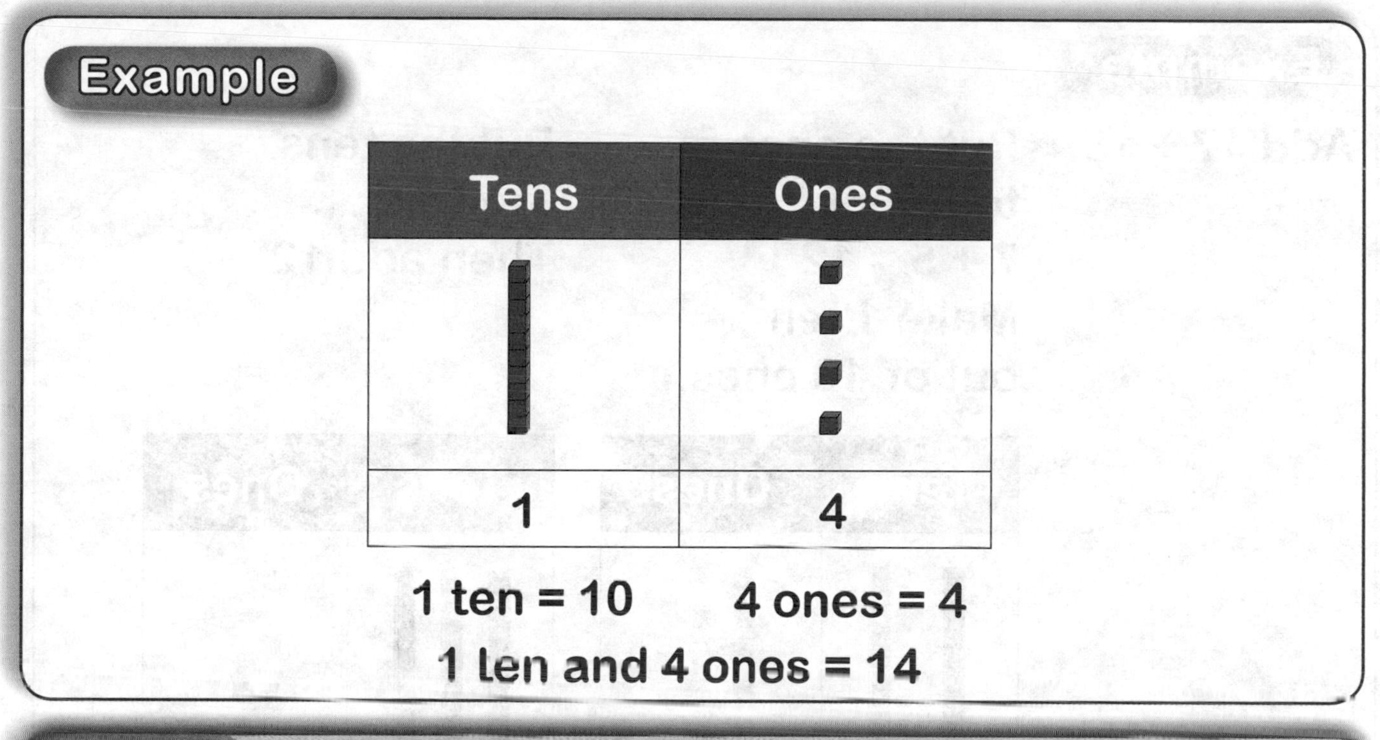

Example

Tens	Ones
1	4

1 ten = 10 4 ones = 4

1 ten and 4 ones = 14

Count

Write how many tens. Write how many ones. Write the number.

1.

Tens	Ones
2	6

The number is 26

2.

Tens	Ones
____	____

The number is ______

Name ____________________

Adding Two-Digit Numbers

You can use models to add.

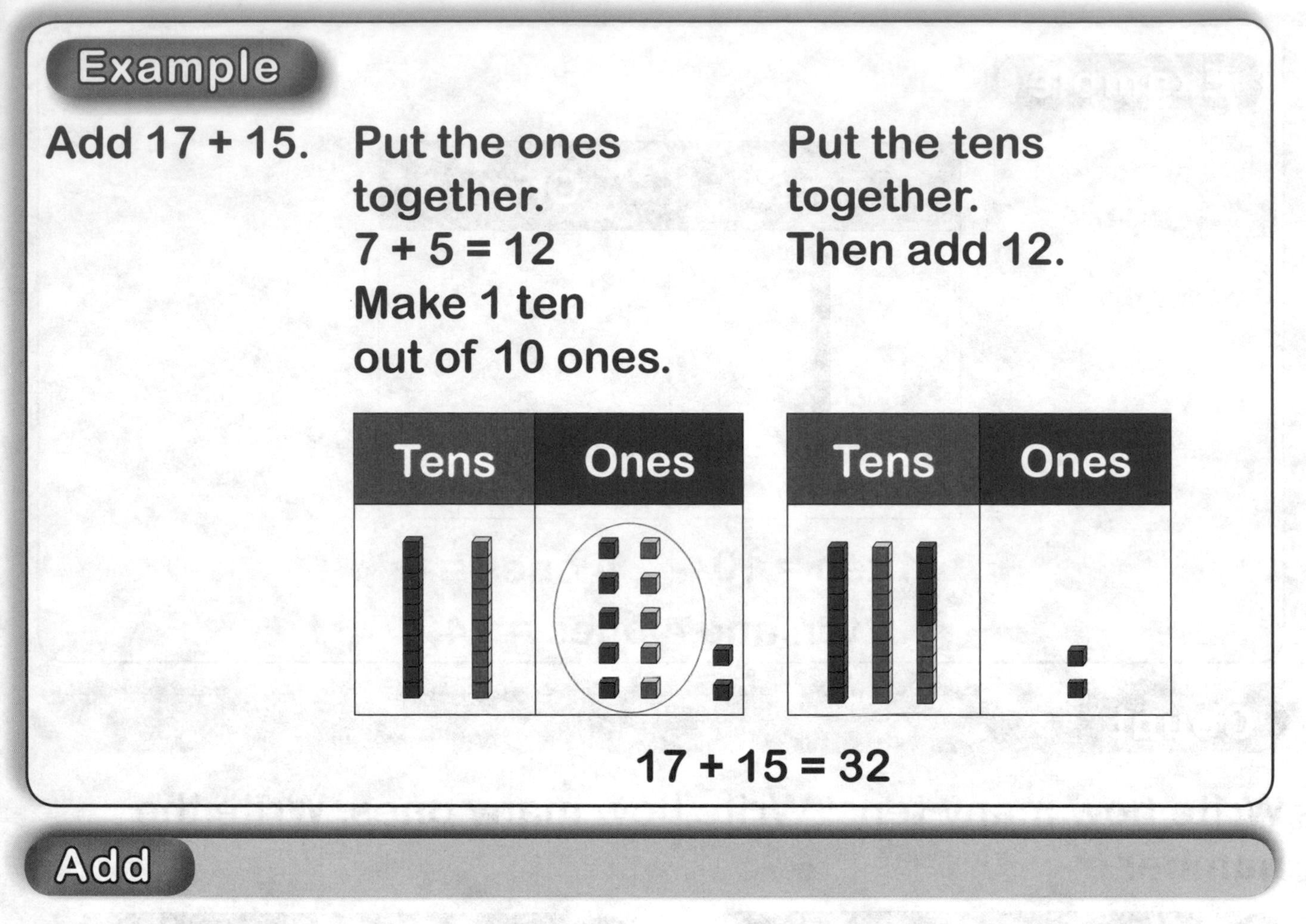

Add

Write the sum.

1. 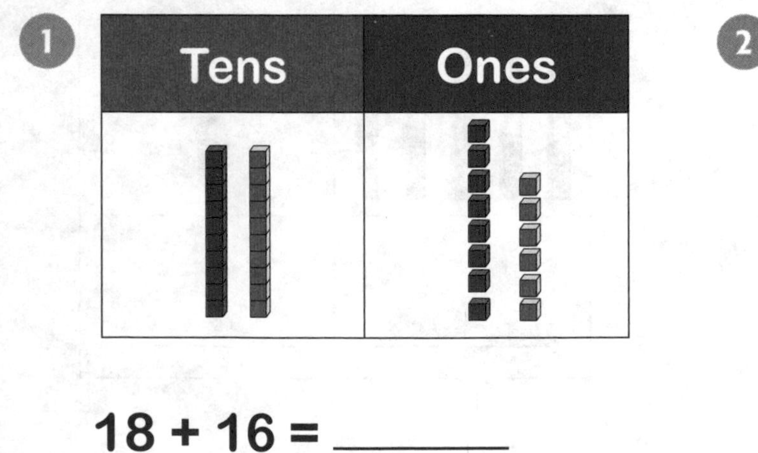

18 + 16 = ______

2. 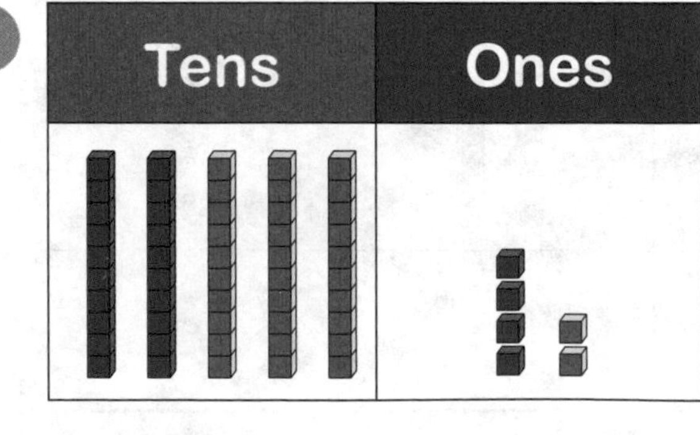

24 + 32 = ______

Name ______________________

Finding 10 More or 10 Less

You can show 10 more than a number. You can show 10 less than a number.

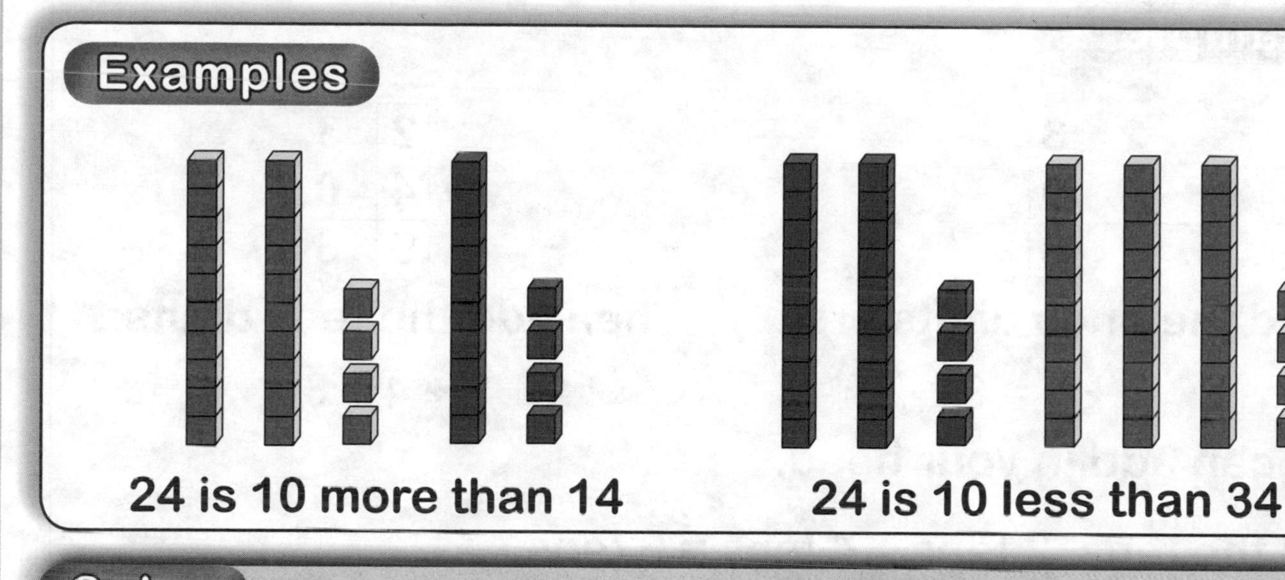

Solve

Find 10 more or 10 less. Write your answer.

1. 42

   10 more than 42 is 52.

   10 less than 42 is 32.

2. 29

   10 more than 29 is ______.

   10 less than 29 is ______.

3. 36

   10 more than 36 is ______.

   10 less than 36 is ______.

Name ______________________

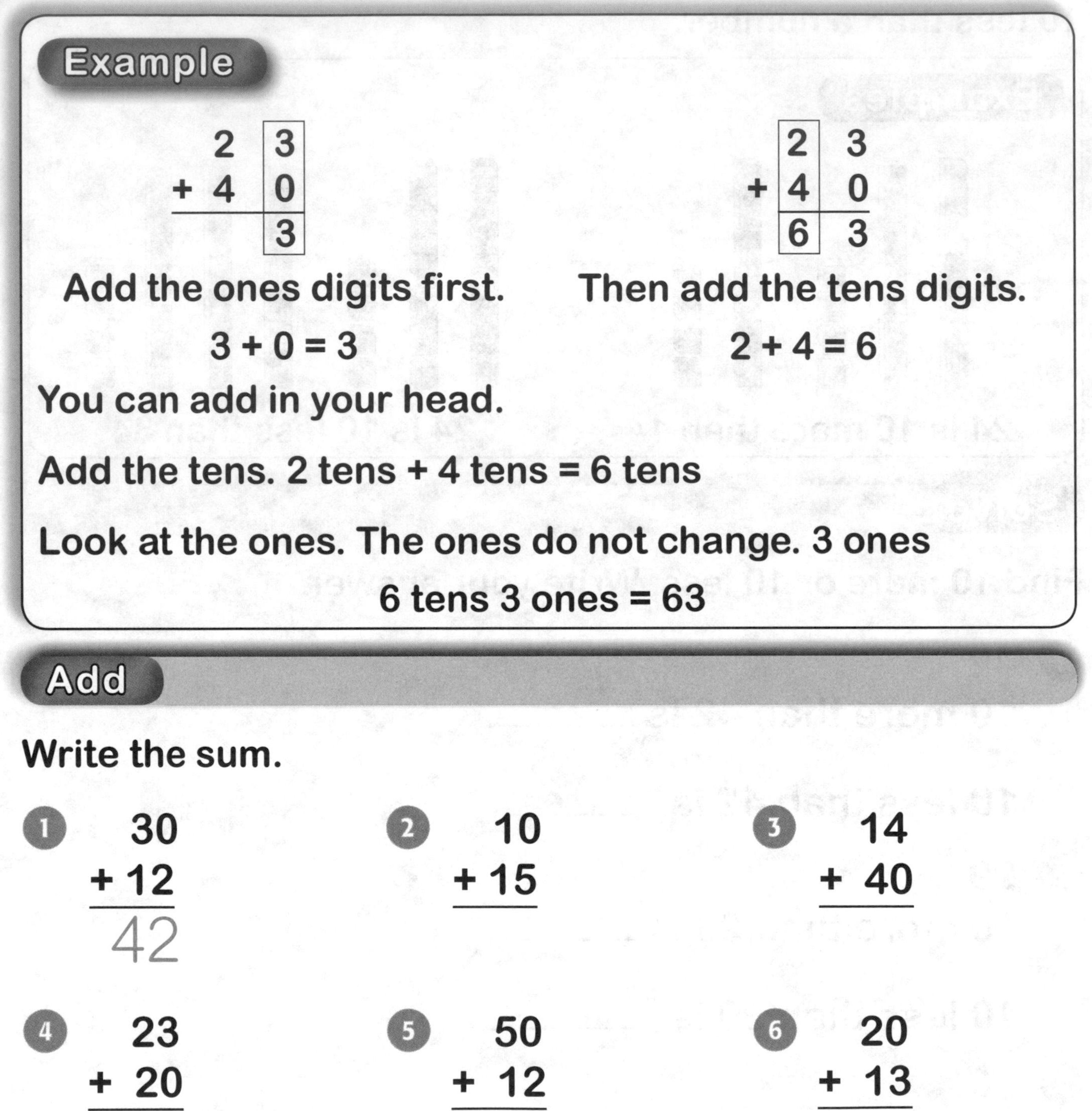

Adding Multiples of 10

You can add tens to a number.

Example

	2	[3]
+	4	[0]
		[3]

Add the ones digits first.

$3 + 0 = 3$

	[2]	3
+	[4]	0
	[6]	3

Then add the tens digits.

$2 + 4 = 6$

You can add in your head.

Add the tens. 2 tens + 4 tens = 6 tens

Look at the ones. The ones do not change. 3 ones

6 tens 3 ones = 63

Add

Write the sum.

1. $30 + 12 = $ 42

2. $10 + 15 = $ ____

3. $14 + 40 = $ ____

4. $23 + 20 = $ ____

5. $50 + 12 = $ ____

6. $20 + 13 = $ ____

Name ______________________

Subtract Multiples of 10

You can subtract multiples of ten.

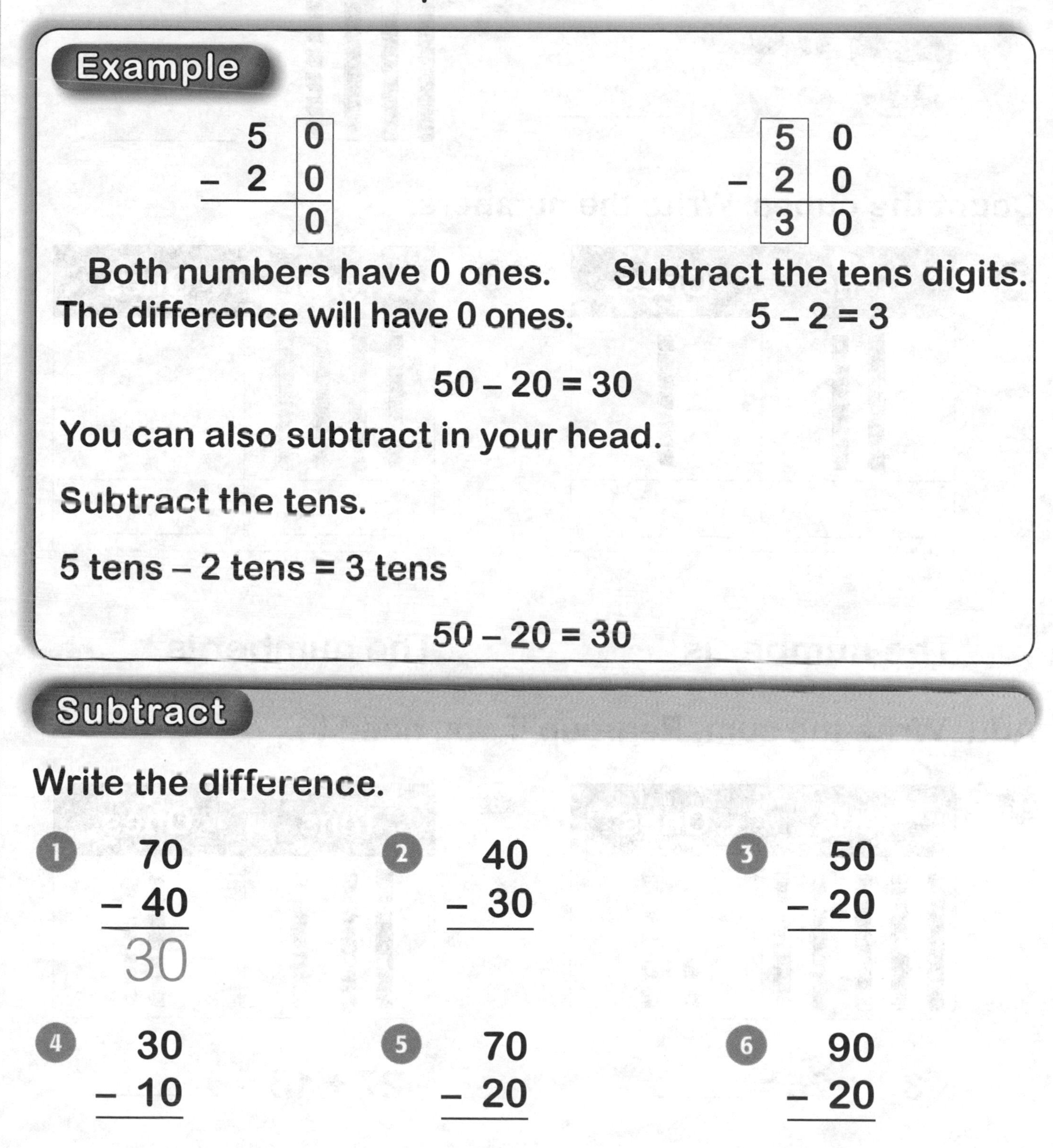

Example

	5	0
−	2	0
		0

Both numbers have 0 ones. The difference will have 0 ones.

	5	0
−	2	0
	3	0

Subtract the tens digits. $5 - 2 = 3$

$$50 - 20 = 30$$

You can also subtract in your head.

Subtract the tens.

5 tens − 2 tens = 3 tens

$$50 - 20 = 30$$

Subtract

Write the difference.

1. $70 - 40 = $ 30
2. $40 - 30 = $ ____
3. $50 - 20 = $ ____
4. $30 - 10 = $ ____
5. $70 - 20 = $ ____
6. $90 - 20 = $ ____

Name ____________________

Look at each picture. Count the tens and ones.
Write the number.

1

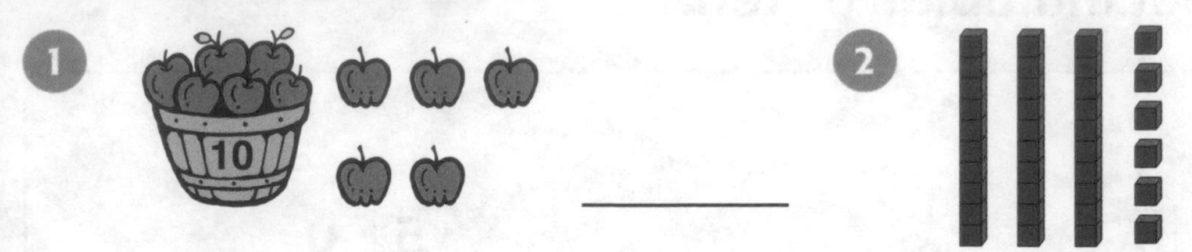

____

2

____

Count the cubes. Write the numbers.

3

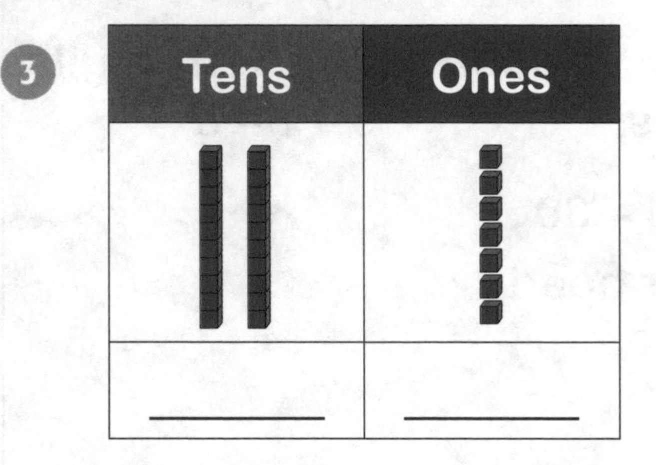

The number is _______

4

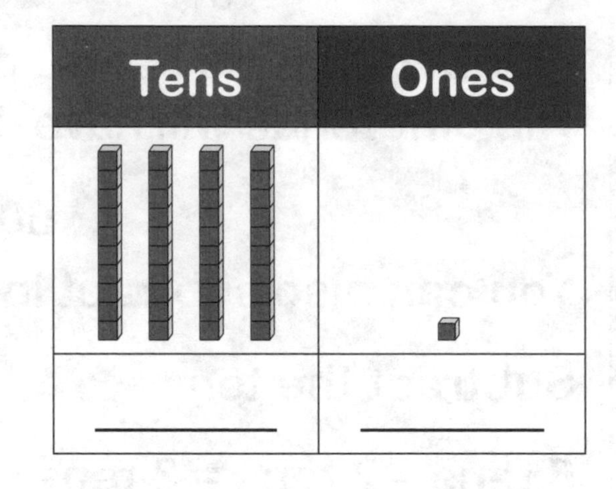

The number is _______

Add. Write the sum. Regroup if you need to.

5

6

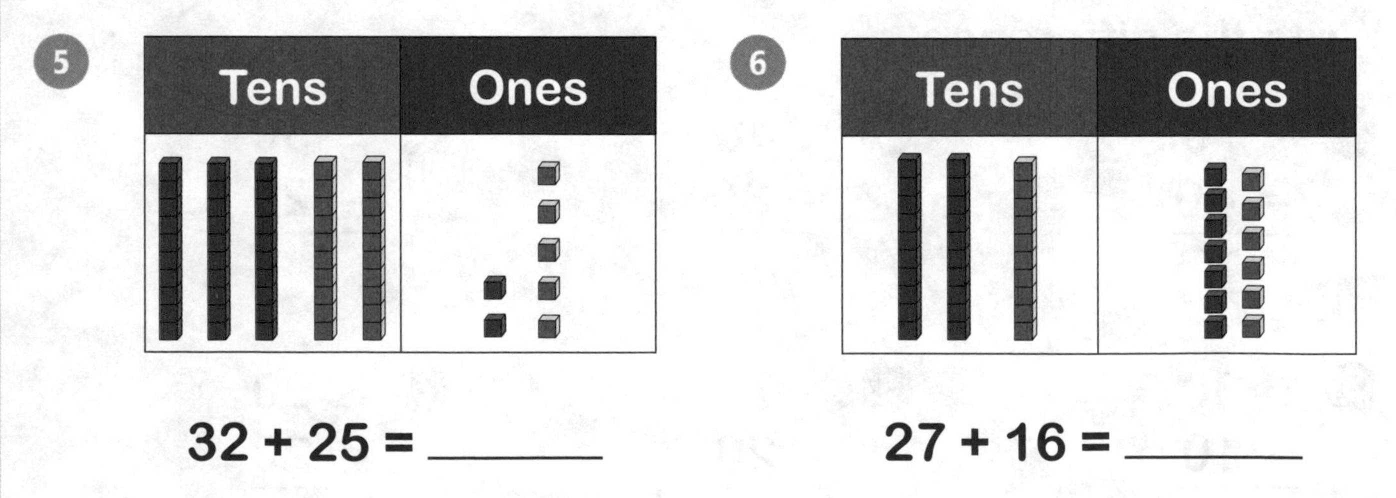

$32 + 25 =$ _______

$27 + 16 =$ _______

Name ______________________

Use objects to count. Write your answer.

7. 24

10 more than 24 is ______

10 less than 24 is ______

8. 81

10 more than 81 is ______

10 less than 81 is ______

Add or subtract. Write the sum or difference.

9. $\begin{array}{r} 35 \\ +\ 4 \\ \hline \end{array}$

10. $\begin{array}{r} 52 \\ +\ 5 \\ \hline \end{array}$

11. $\begin{array}{r} 23 \\ +\ 60 \\ \hline \end{array}$

12. $\begin{array}{r} 36 \\ -\ 30 \\ \hline \end{array}$

13. $\begin{array}{r} 30 \\ -\ 20 \\ \hline \end{array}$

14. $\begin{array}{r} 70 \\ -\ 50 \\ \hline \end{array}$

15. $\begin{array}{r} 40 \\ +\ 30 \\ \hline \end{array}$

16. $\begin{array}{r} 28 \\ +\ 20 \\ \hline \end{array}$

17. $\begin{array}{r} 18 \\ +\ 50 \\ \hline \end{array}$

Name ___________________________

Length and Weight

You measure to find the amount or size of an object. Length tells how long. Weight tells how heavy.

Examples

Length

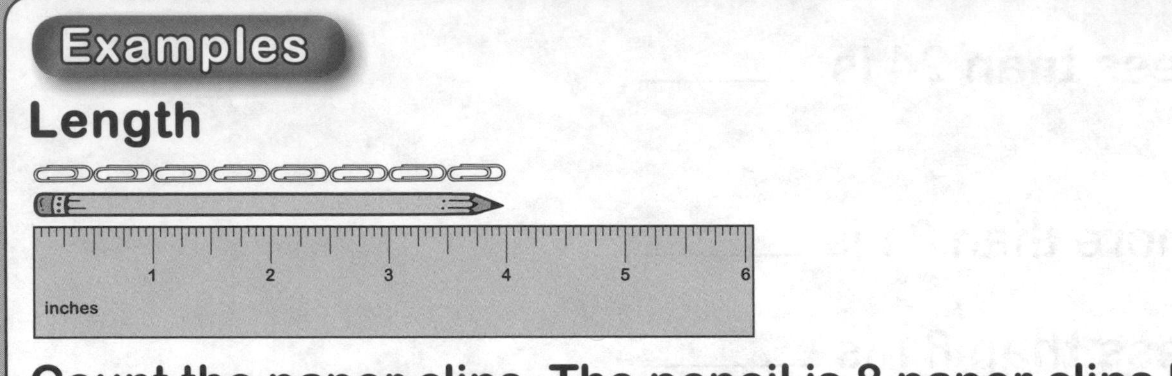

Count the paper clips. The pencil is 8 paper clips long.

The pencil is 4 inches long.

Weight

A scale can show which is heavier.

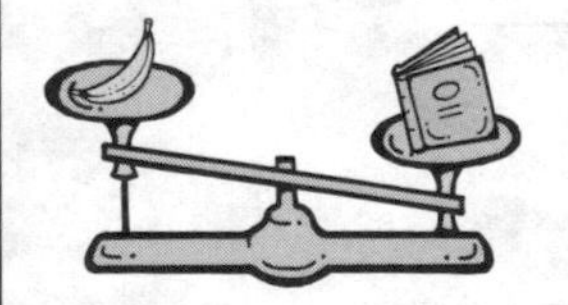

The book is heavier than the banana.

Measure

Write your answer.

1. Find a pen. How long is the pen? Use paper clips.
   The pen is __________ paper clips long.

Circle the object that is heavier.

2.  

3.  

Name ______________________________

Ways to Measure

You can measure the same object in different ways.

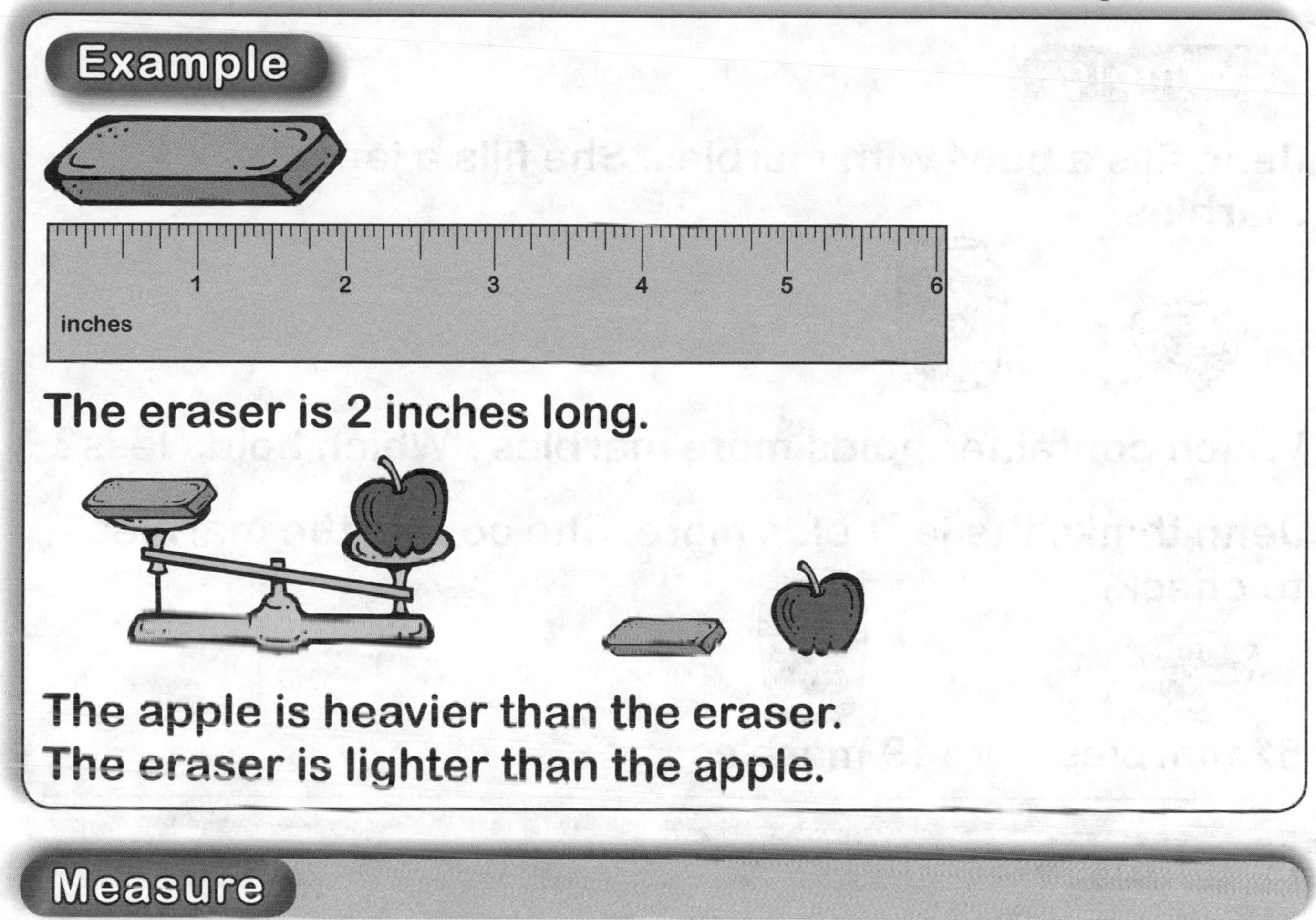

The eraser is 2 inches long.

The apple is heavier than the eraser.
The eraser is lighter than the apple.

Measure

Write or draw your answers.

1. Find a spoon. How long is the spoon? Use a ruler.
   The spoon is __________ inches long.

2. 

   Which object is heavier? Circle your answer.

Name ____________________

"More" or "Less"

Some objects hold more. Some objects hold less.

Example

Jenn fills a bowl with marbles. She fills a jar with marbles.

Which container holds more marbles? Which holds less?

Jenn thinks the jar holds more. She counts the marbles to check.

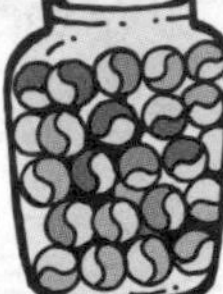

52 marbles   119 marbles

Compare

Circle the container you think holds more.

Name ____________________

Order Three Objects by Length

You can put objects in order from longest to shortest.

Example

The red spoon is longer than the blue spoon and the yellow spoon.

The blue spoon is shorter than the red spoon and the yellow spoon.

Compare

1. Which crayon is longest? Color it blue.
2. Which crayon is shortest? Color it green.

3. Color the shortest pencil red.
4. Color the longest pencil blue. Make the other pencil orange.

Name ______________________

Comparing Lengths of Objects

You can compare lengths of objects.

Examples

The pen is shorter than the eraser.

The pen is longer than the spoon.

Compare

Measure the objects in the picture. You can use string, paper clips, or other objects to measure. Which is longer? Circle the longer object.

1
2

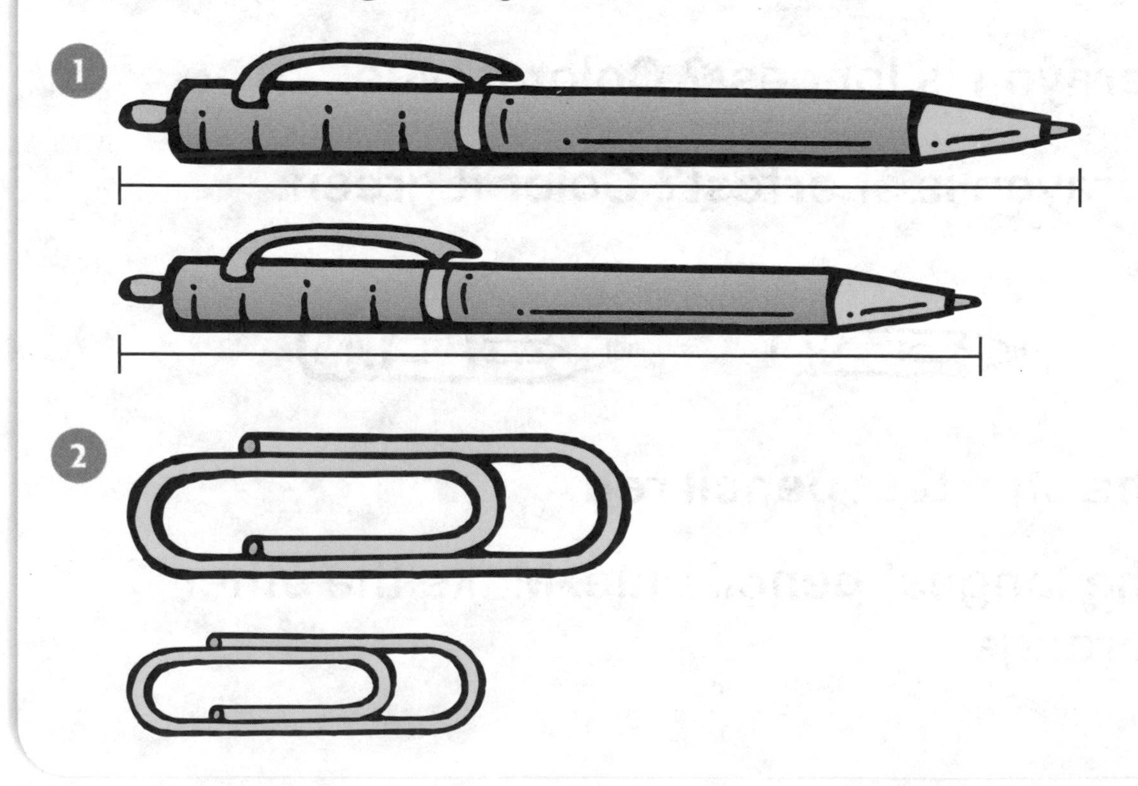

Name ____________________

Sorting Objects

You can put matching objects into groups.
This is called sorting.

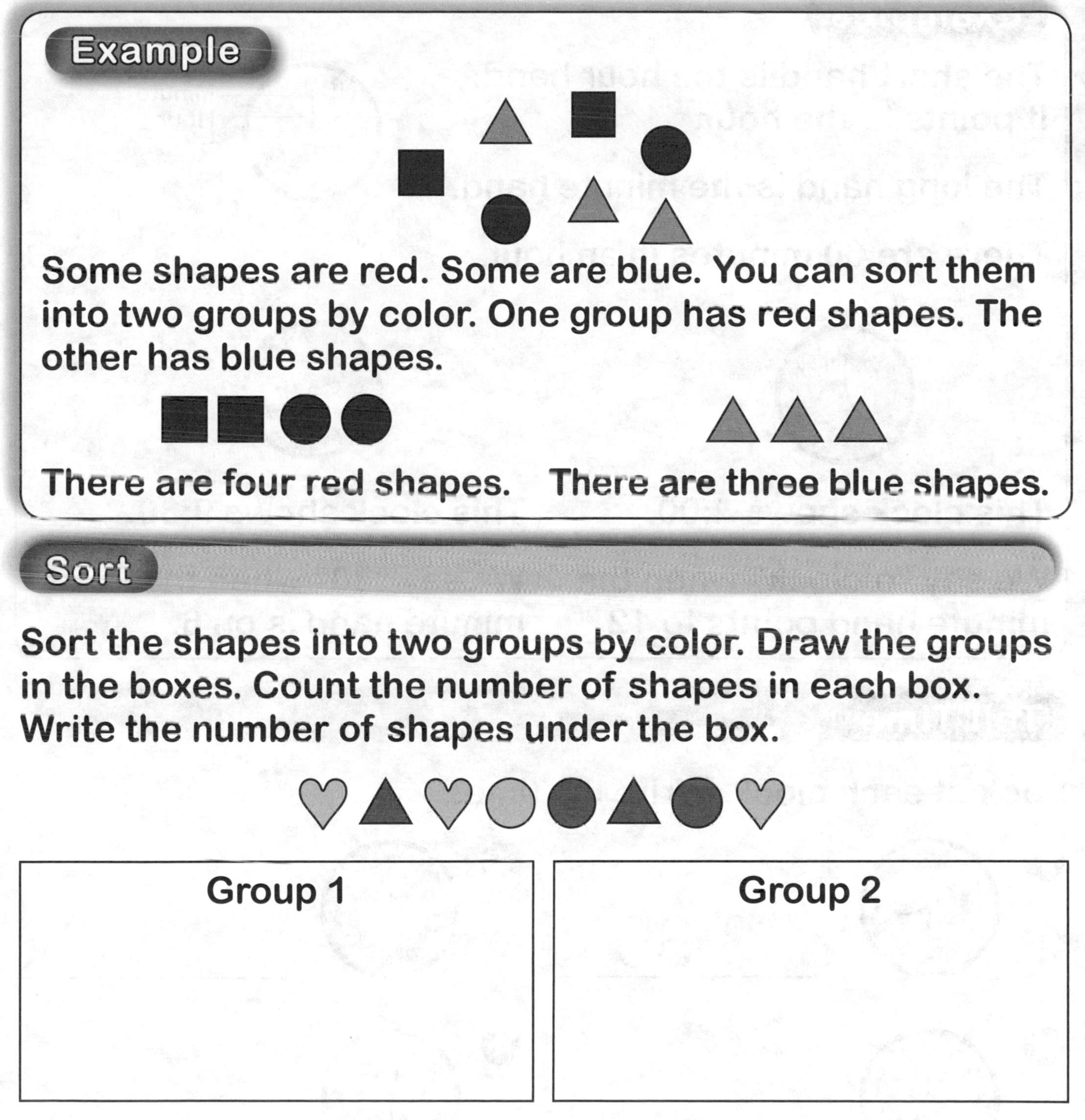

Example

Some shapes are red. Some are blue. You can sort them into two groups by color. One group has red shapes. The other has blue shapes.

There are four red shapes. There are three blue shapes.

Sort

Sort the shapes into two groups by color. Draw the groups in the boxes. Count the number of shapes in each box. Write the number of shapes under the box.

Group 1	Group 2

__________ __________

Name ____________________

Time in Hours and Half-Hours

You can use a clock to tell time.

Examples

The short hand is the hour hand.
It points to the hour.

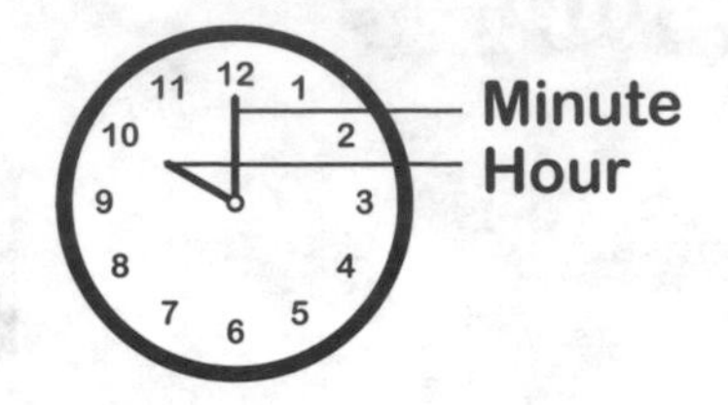

The long hand is the minute hand.

There are 60 minutes in an hour.

This clock shows 4:00.

We say "o'clock" when the minute hand points to 12.

This clock shows 9:30.

We say "30" when the minute hand is on 6.

Tell Time

Look at each clock. Write the time.

Name ______________________________

Telling Time with Digital Clocks

A digital clock does not use minute or hour hands. It shows time with numbers and a :

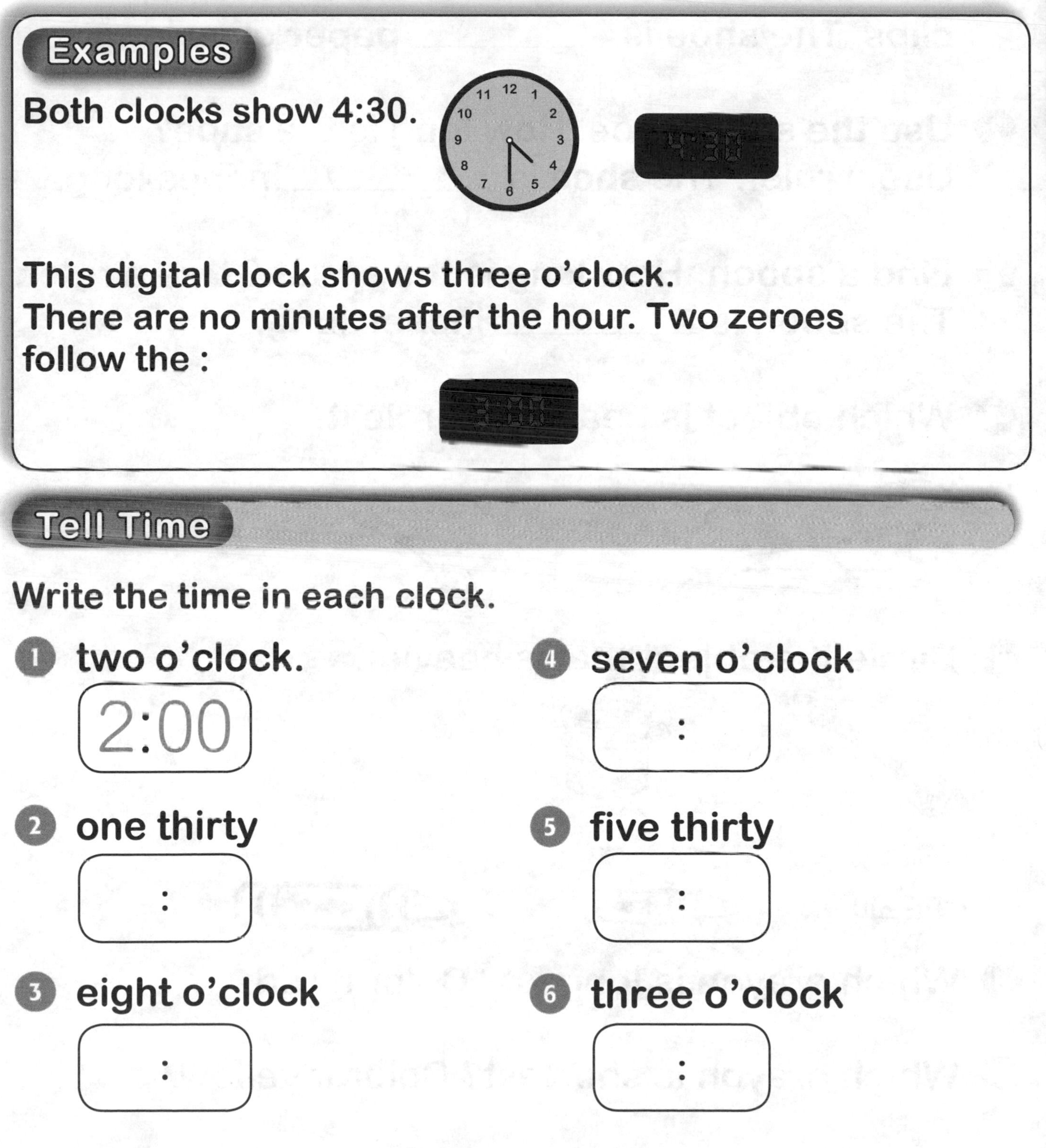

Examples

Both clocks show 4:30.

This digital clock shows three o'clock. There are no minutes after the hour. Two zeroes follow the :

Tell Time

Write the time in each clock.

1. two o'clock. 2:00
2. one thirty ___ : ___
3. eight o'clock ___ : ___
4. seven o'clock ___ : ___
5. five thirty ___ : ___
6. three o'clock ___ : ___

Name ______________________

Use paper clips or other objects and a ruler to measure. Write your answers.

1. Find a shoe. How long is the shoe? Use paper clips. The shoe is __________ paper clips long.

2. Use the same shoe. How long is the shoe? Use a ruler. The shoe is __________ inches long.

3. Find a spoon. How long is the spoon? Use a ruler. The spoon is __________ inches long.

4. Which object is heavier? Circle it.

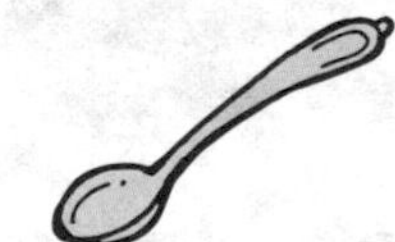

5. Circle the object that is heavier.

6. Which crayon is longest? Color it red.

7. Which crayon is shortest? Color it yellow.

Name ______________________

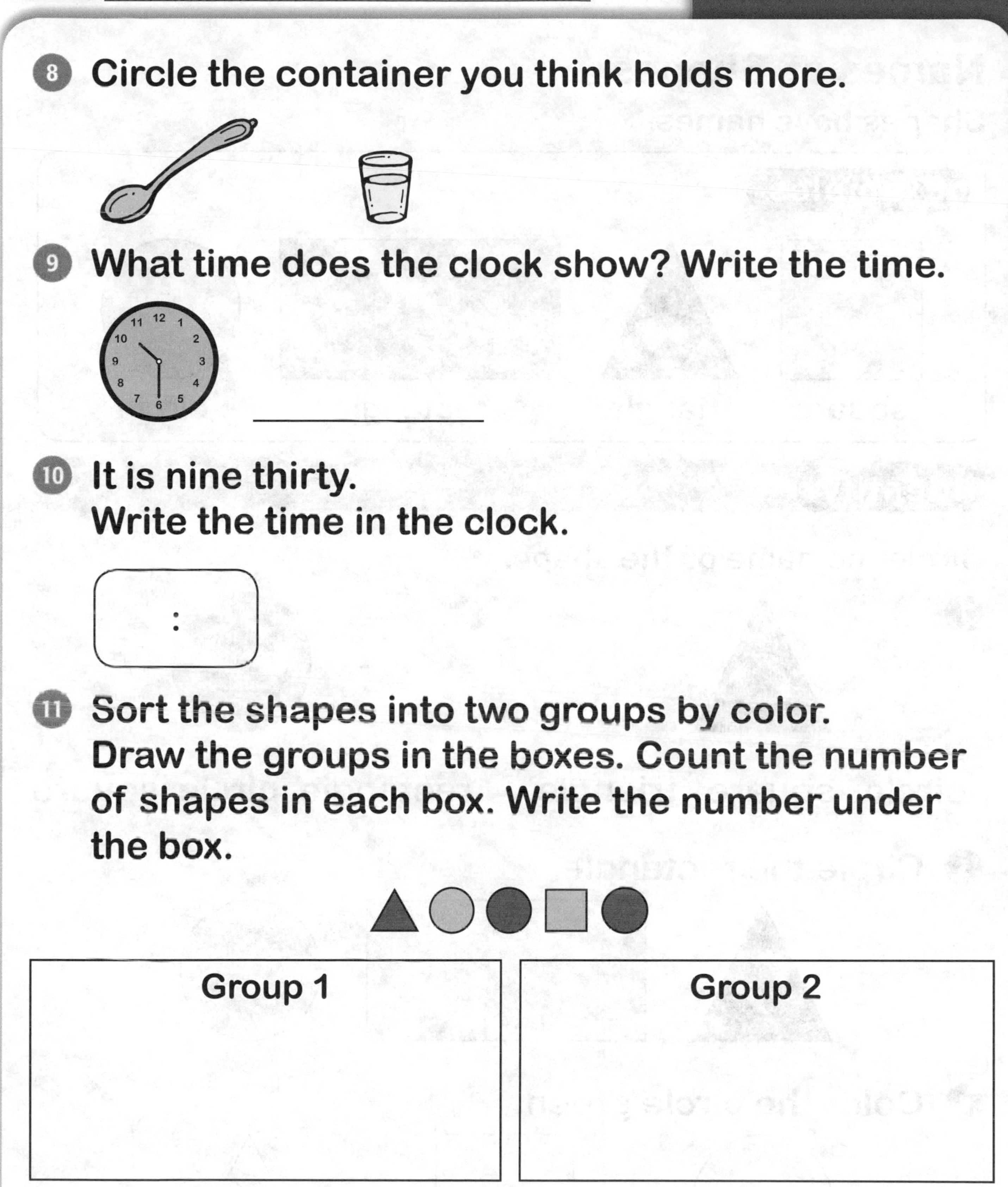

8 Circle the container you think holds more.

9 What time does the clock show? Write the time.

____

10 It is nine thirty.
Write the time in the clock.

11 Sort the shapes into two groups by color.
Draw the groups in the boxes. Count the number of shapes in each box. Write the number under the box.

Group 1	Group 2

______________ ______________

Name ______________________

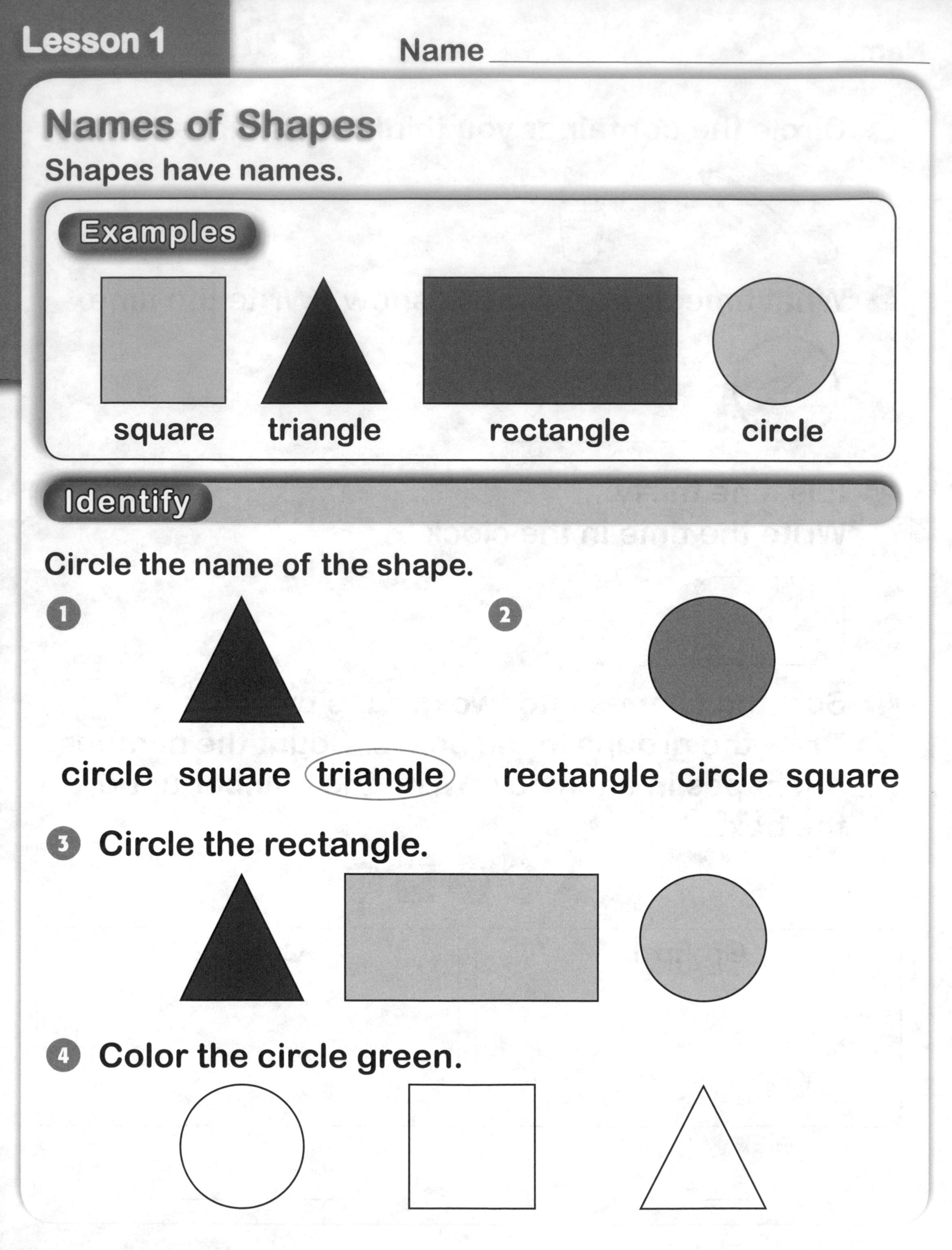

Names of Shapes

Shapes have names.

Examples

square triangle rectangle circle

Identify

Circle the name of the shape.

1. circle square triangle

2. rectangle circle square

3. Circle the rectangle.

4. Color the circle green.

Name ______________________________

Shapes Can Be Different Colors

Shapes can be different colors. They keep their names.

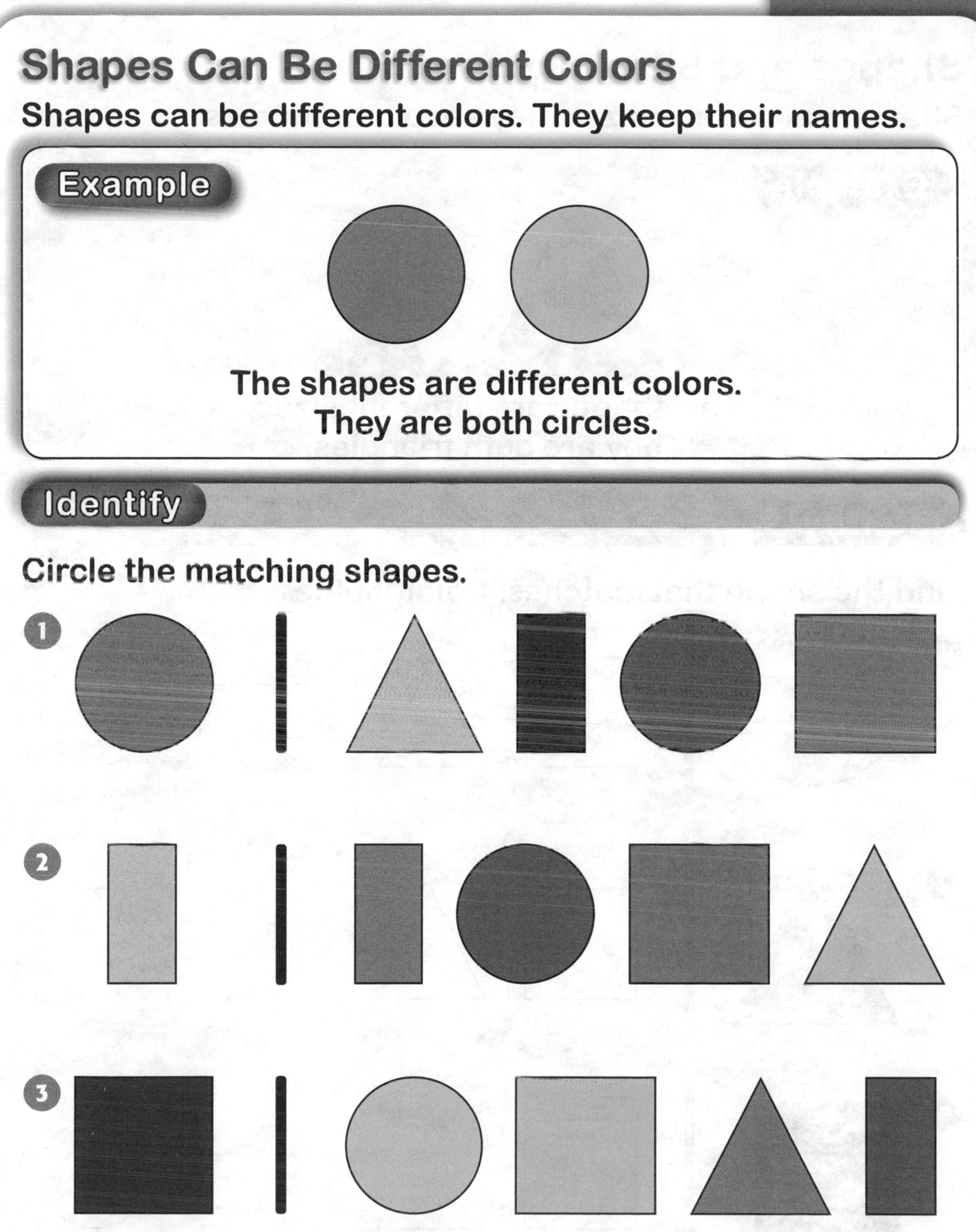

Example

The shapes are different colors.
They are both circles.

Identify

Circle the matching shapes.

1

2

3

Name ________________________

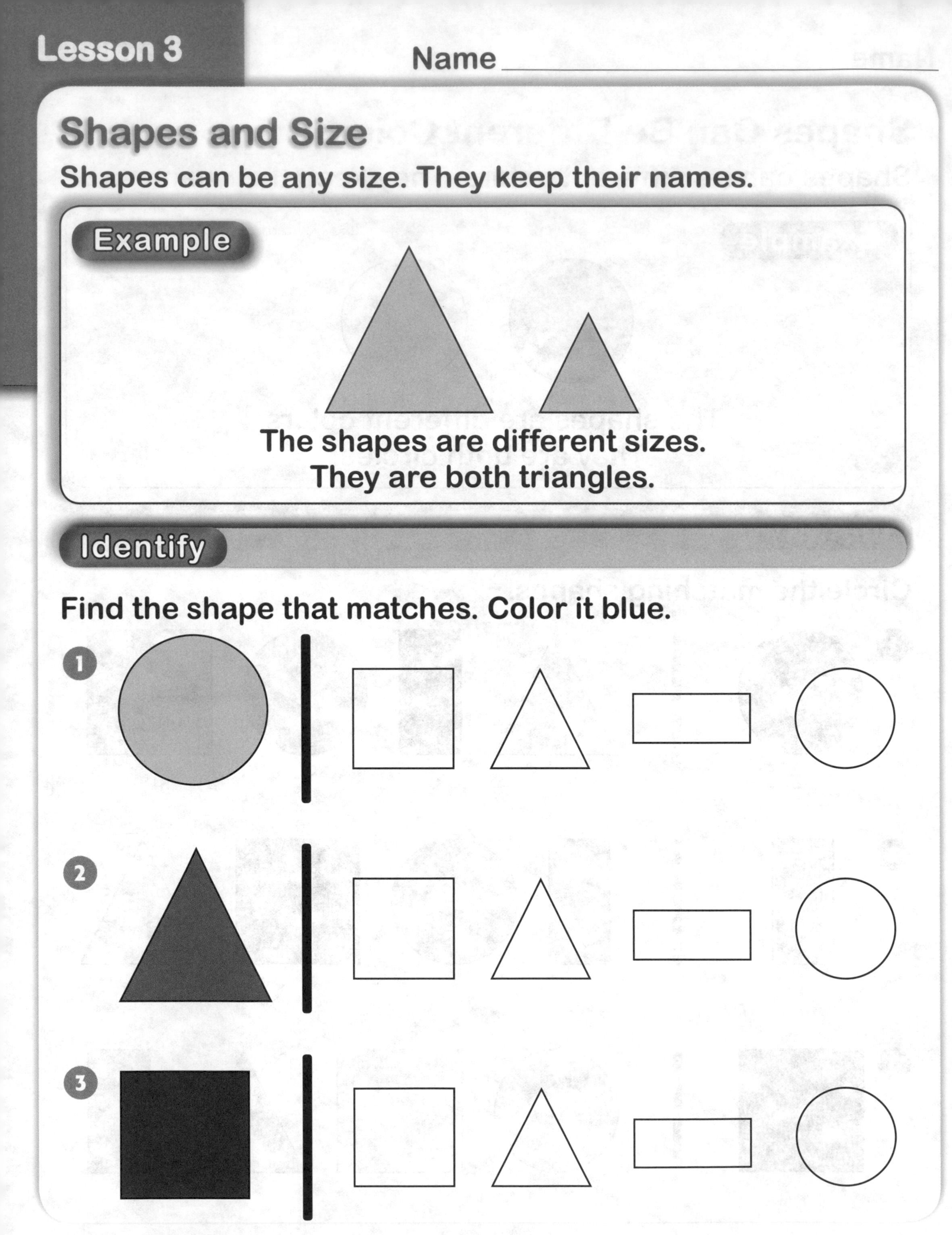

Shapes and Size

Shapes can be any size. They keep their names.

Example

The shapes are different sizes.
They are both triangles.

Identify

Find the shape that matches. Color it blue.

1

2

3

Name ______________________

Turned Shapes

Shapes can be turned. They still keep their names.

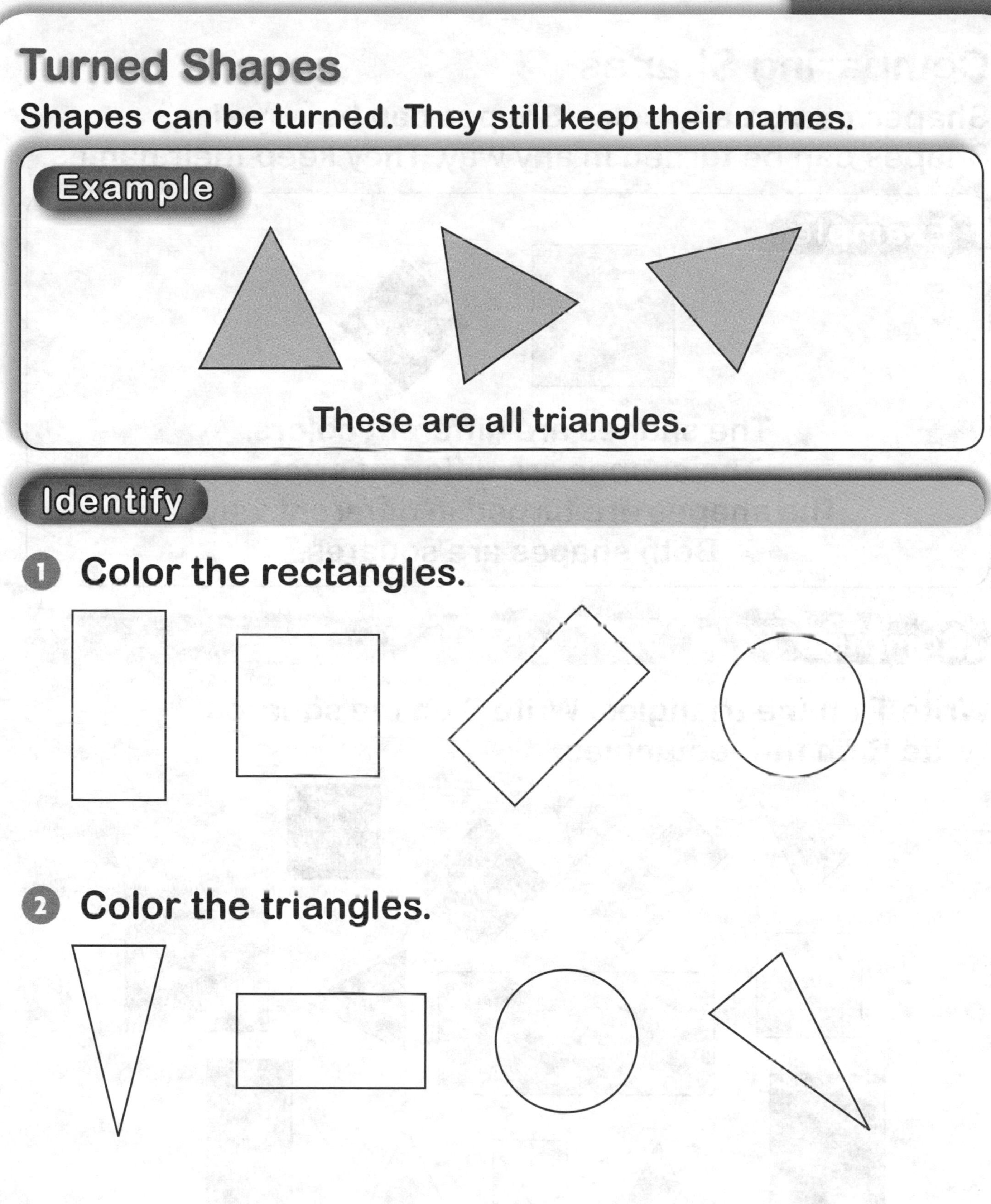

1. Color the rectangles.

2. Color the triangles.

Name ___________________________

Comparing Shapes

Shapes can be any color. Shapes can be any size.
Shapes can be turned in any way. They keep their names.

Example

The shapes are different colors.
The shapes are different sizes.
The shapes are turned in different ways.
Both shapes are squares.

Compare

Write T on the triangles. Write S on the squares.
Write R on the rectangles.

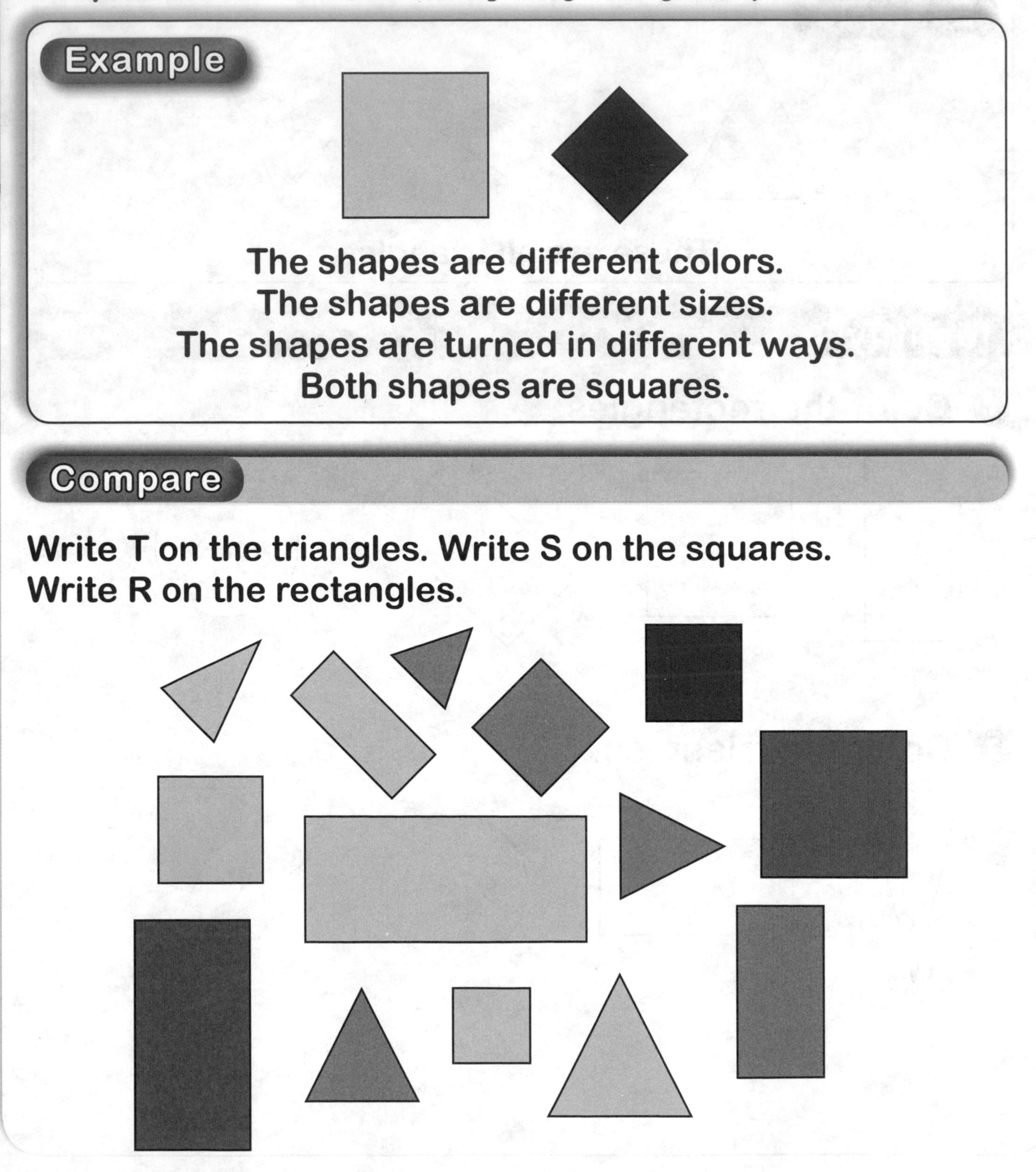

Name ______________________________

Drawing Shapes

You can draw shapes. Dot paper can make it easier.

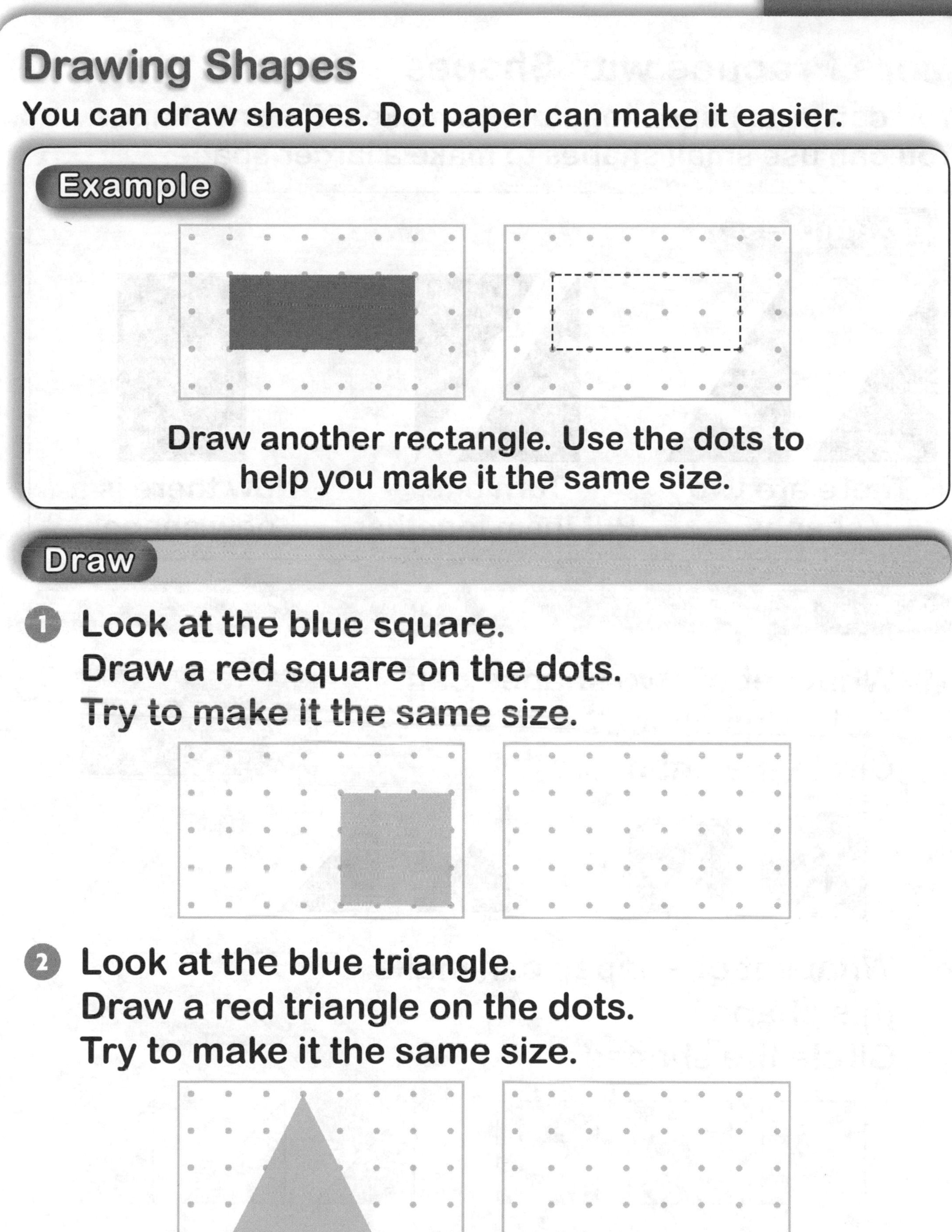

Example

Draw another rectangle. Use the dots to help you make it the same size.

Draw

1. Look at the blue square.
   Draw a red square on the dots.
   Try to make it the same size.

2. Look at the blue triangle.
   Draw a red triangle on the dots.
   Try to make it the same size.

Name ____________________

More Practice with Shapes

You can put shapes together to make different shapes.
You can use small shapes to make a larger shape.

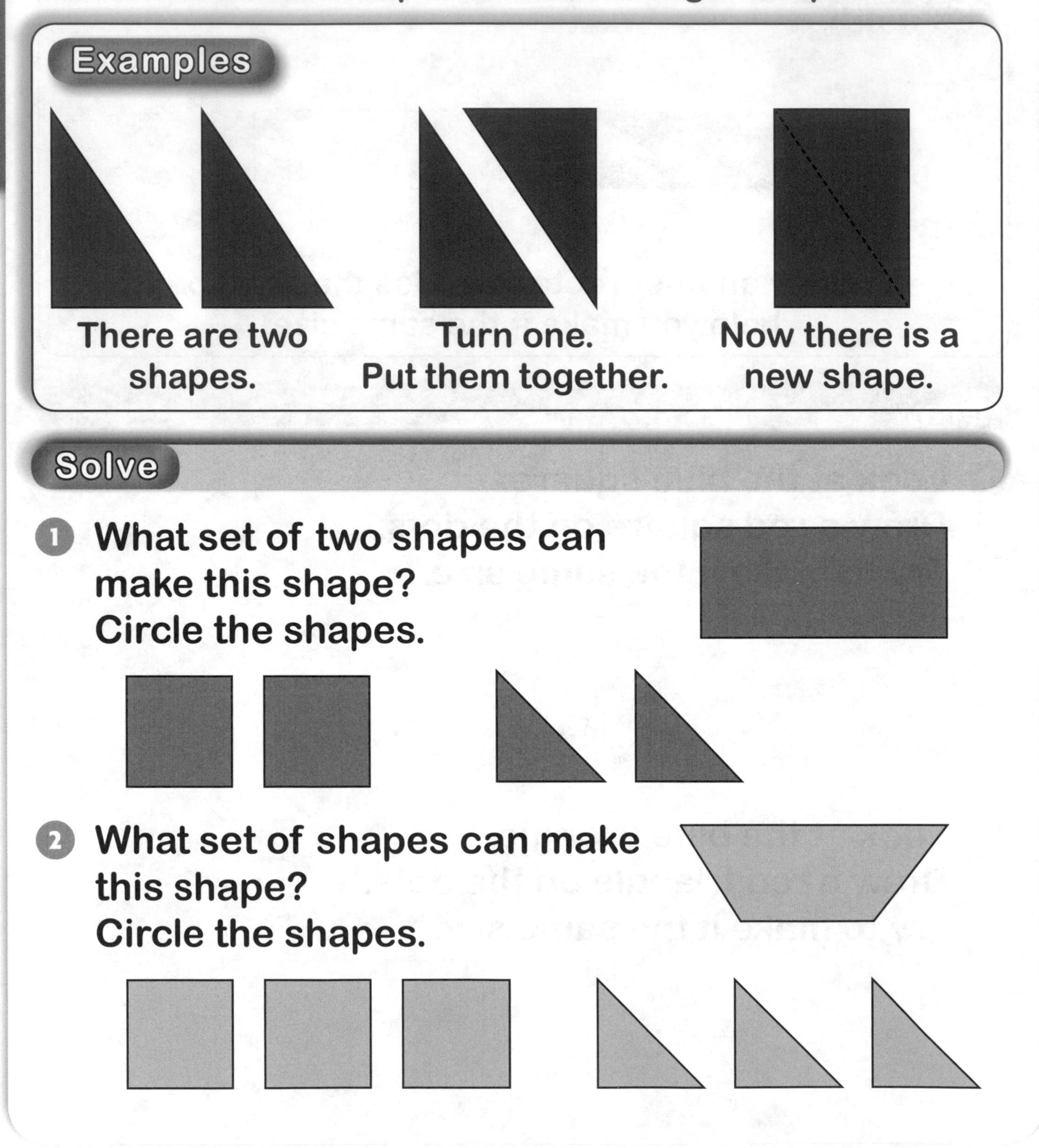

Examples

There are two shapes.

Turn one. Put them together.

Now there is a new shape.

Solve

1. What set of two shapes can make this shape? Circle the shapes.

2. What set of shapes can make this shape? Circle the shapes.

Name ____________________

Shapes in Objects

Solid objects are not flat.
Some objects have sides that look like flat shapes.

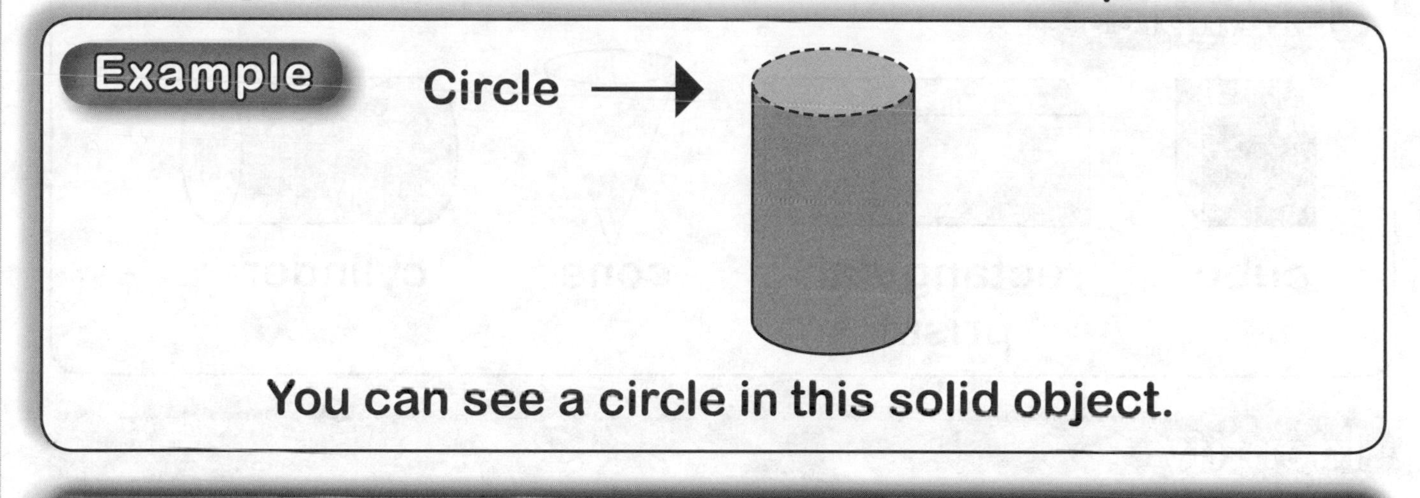

Identify

Look at each object. What shape do you see?
Circle the name.

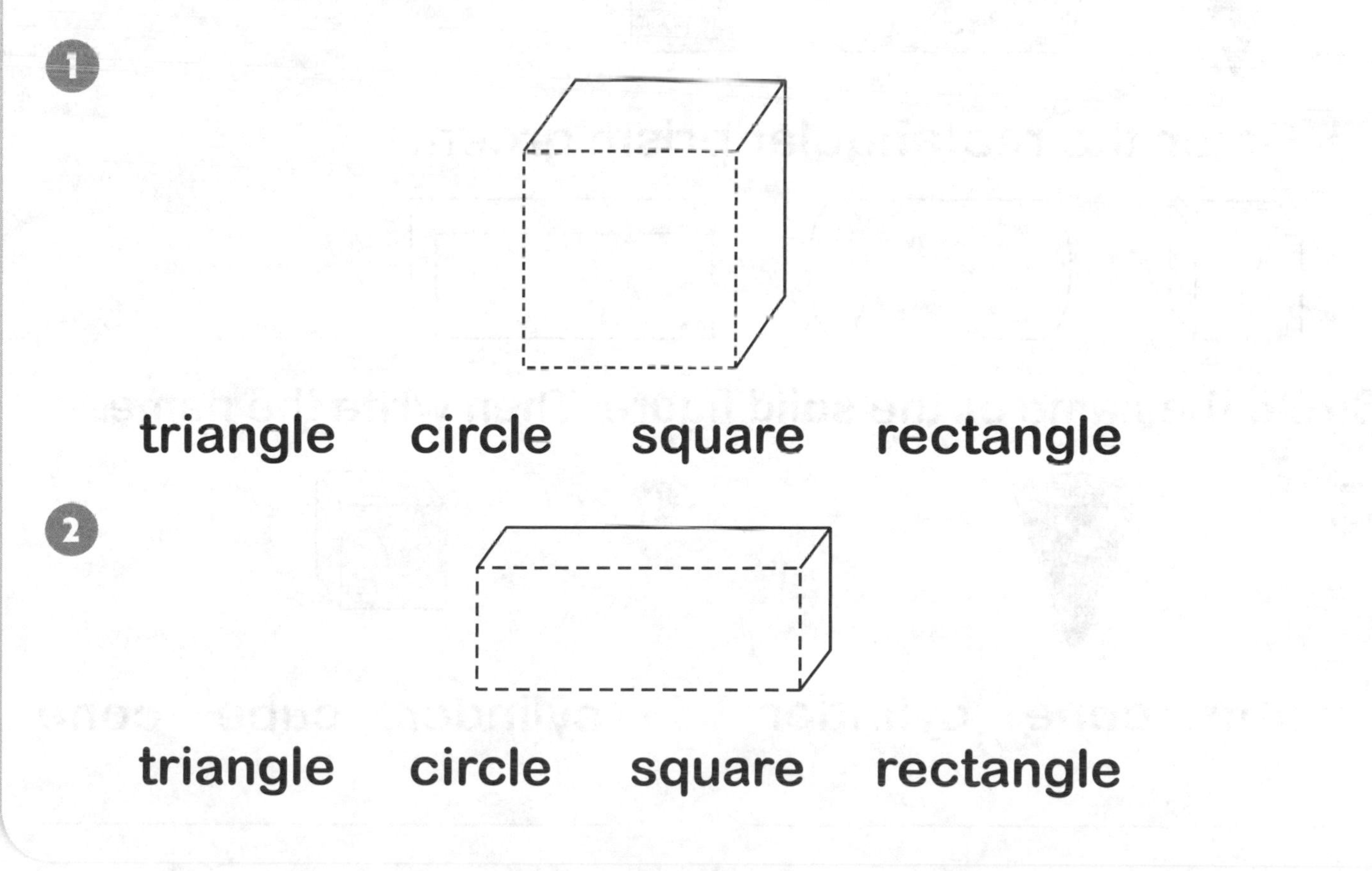

Name ______________________________

Solid Figures

Solid figures have length, width, and height.

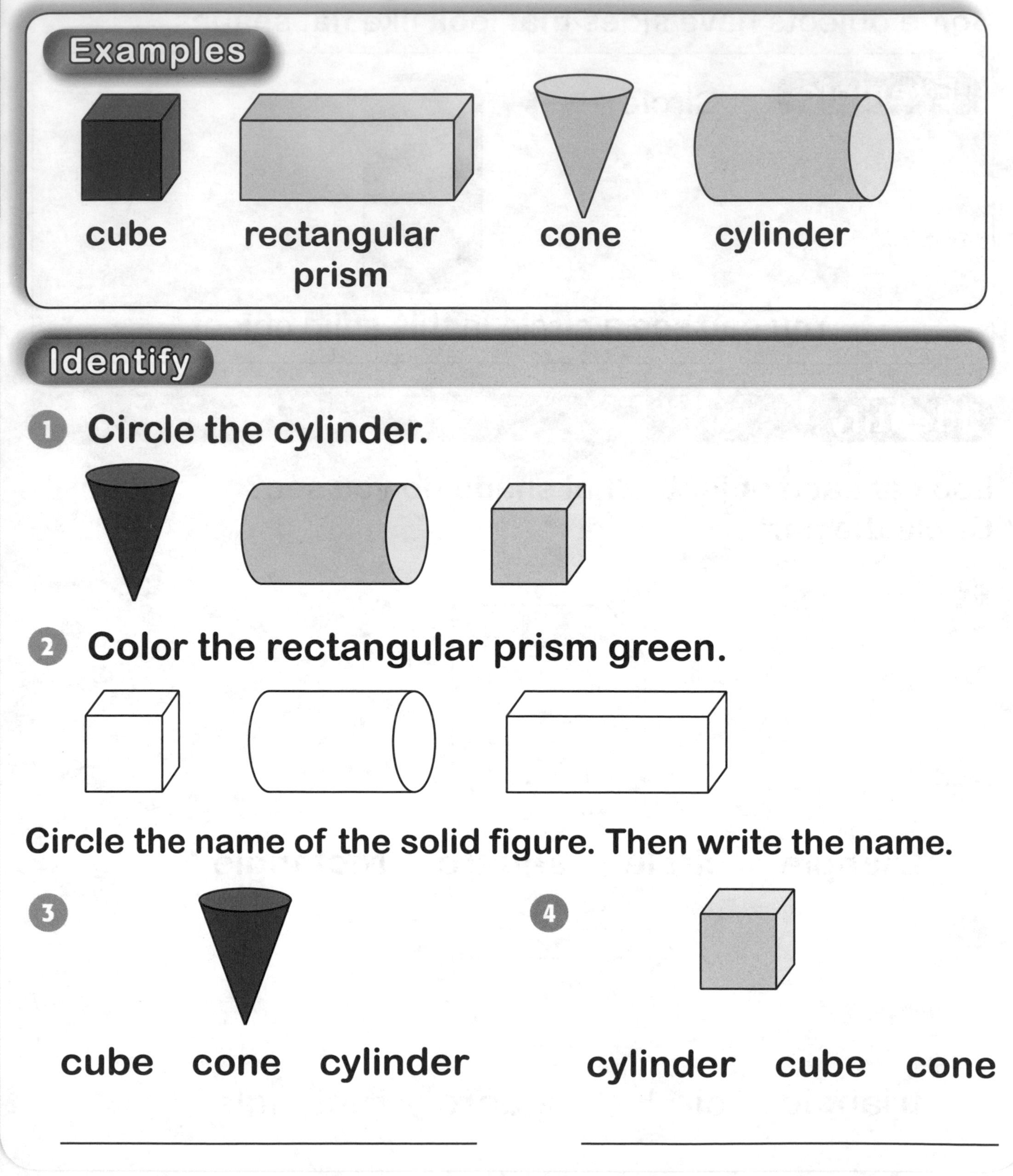

Examples

cube

rectangular prism

cone

cylinder

Identify

1. Circle the cylinder.

2. Color the rectangular prism green.

Circle the name of the solid figure. Then write the name.

3. cube cone cylinder

____

4. cylinder cube cone

____

Name ______________________________

Finding Solid Figures

Many objects look like solid figures.

Example

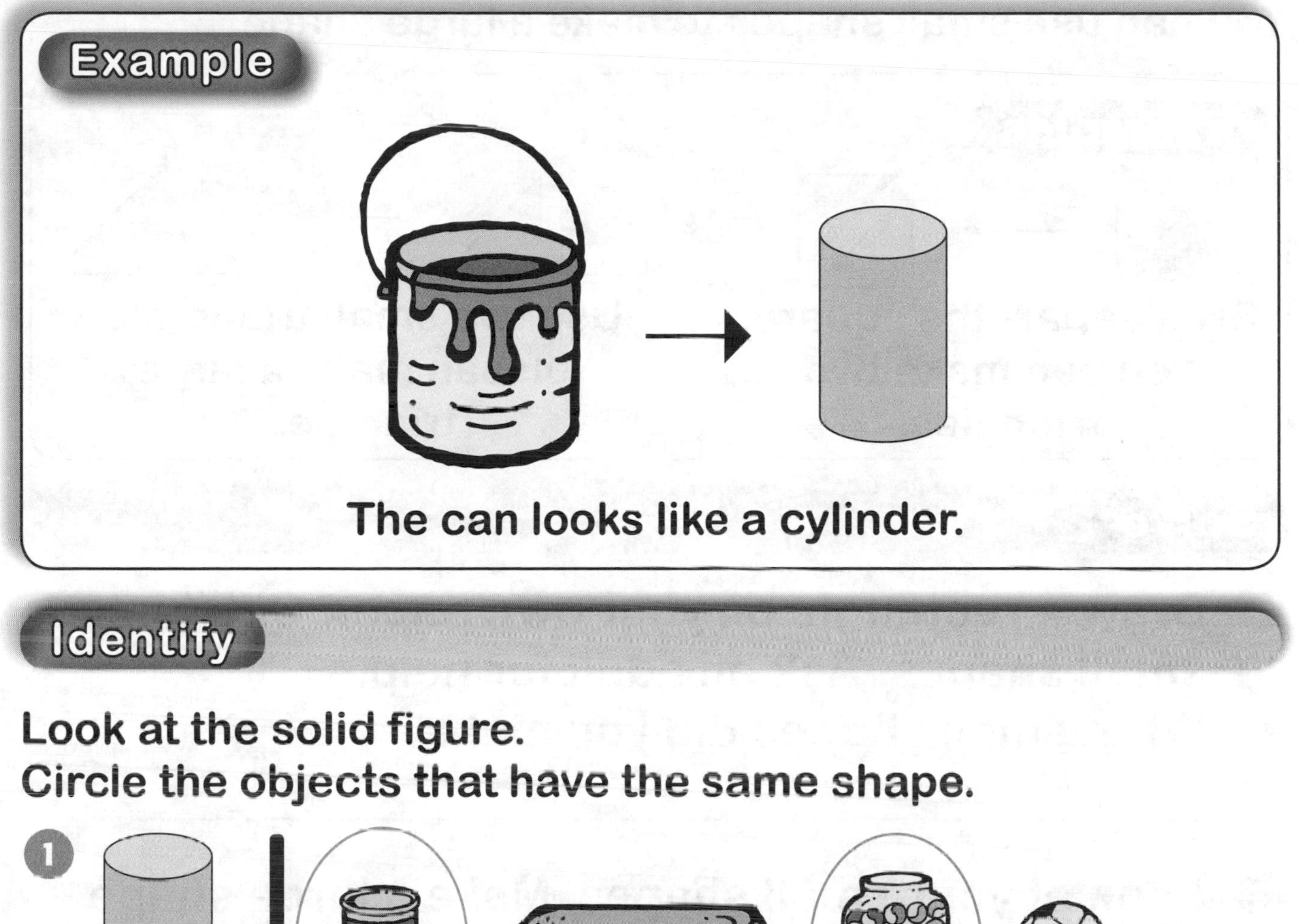

The can looks like a cylinder.

Identify

Look at the solid figure.
Circle the objects that have the same shape.

Lesson 11

Name ______________________

Making Shapes

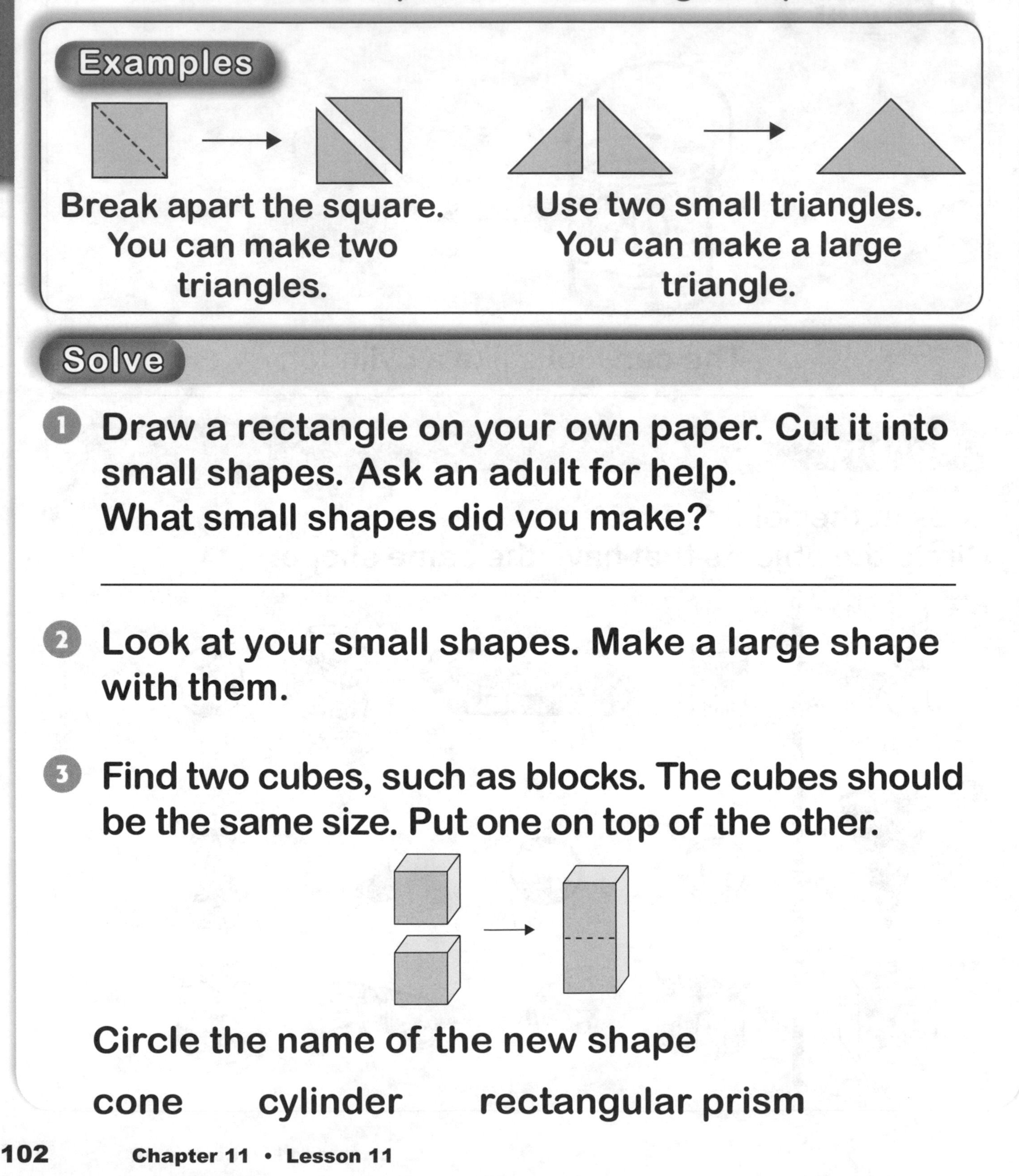

You can break apart large shapes to make small shapes.
You can use small shapes to make a large shape.

Examples

Break apart the square. You can make two triangles.

Use two small triangles. You can make a large triangle.

Solve

1. Draw a rectangle on your own paper. Cut it into small shapes. Ask an adult for help.
   What small shapes did you make?

   ____

2. Look at your small shapes. Make a large shape with them.

3. Find two cubes, such as blocks. The cubes should be the same size. Put one on top of the other.

   Circle the name of the new shape

   cone   cylinder   rectangular prism

Name ____________________

Making Equal Parts

You can divide shapes into equal parts.

Examples

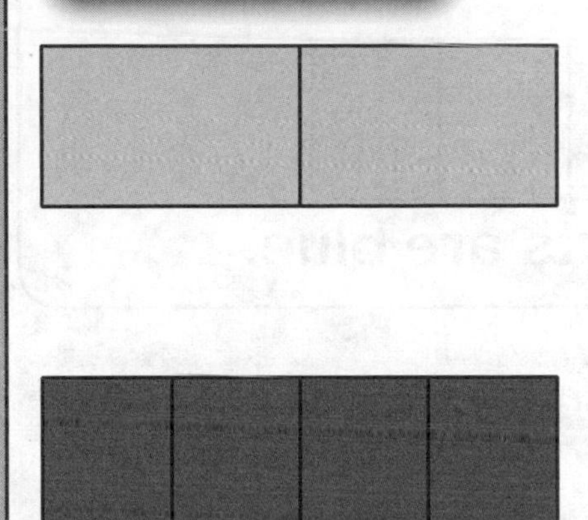

A shape can be divided into 2 equal parts. Each equal part is a half of the shape. This shape has 2 halves.

A shape can be divided into 4 equal parts. Each equal part is a fourth of the shape. Each equal part is also called a quarter. This shape has 4 quarters.

Identify

1. How many equal parts does the shape have? 4

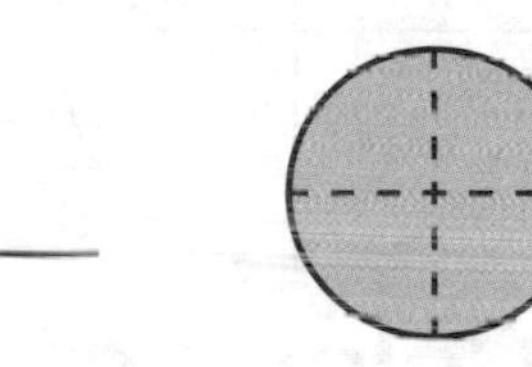

Is the shape divided into halves or fourths? ____________

2. Divide the circle into halves. How many equal parts does the circle have? ______

3. Divide the rectangle into fourths. How many equal parts does the rectangle have? ______

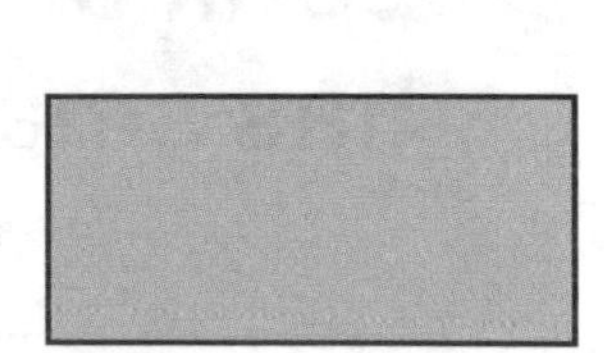

Name ______________________

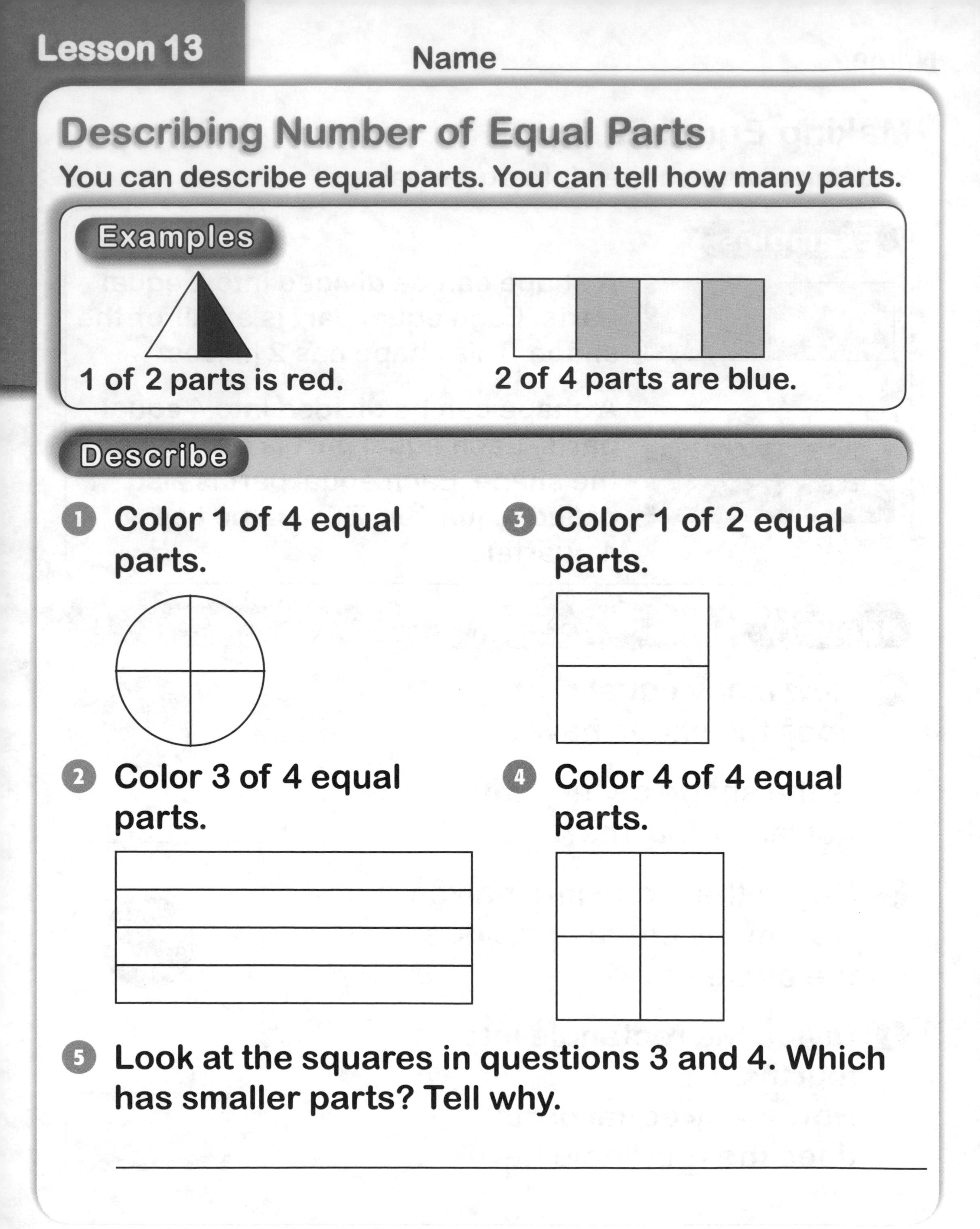

Describing Number of Equal Parts

You can describe equal parts. You can tell how many parts.

Examples

1 of 2 parts is red.

2 of 4 parts are blue.

Describe

1. Color 1 of 4 equal parts.

3. Color 1 of 2 equal parts.

2. Color 3 of 4 equal parts.

4. Color 4 of 4 equal parts.

5. Look at the squares in questions 3 and 4. Which has smaller parts? Tell why.

____

Name ______________________

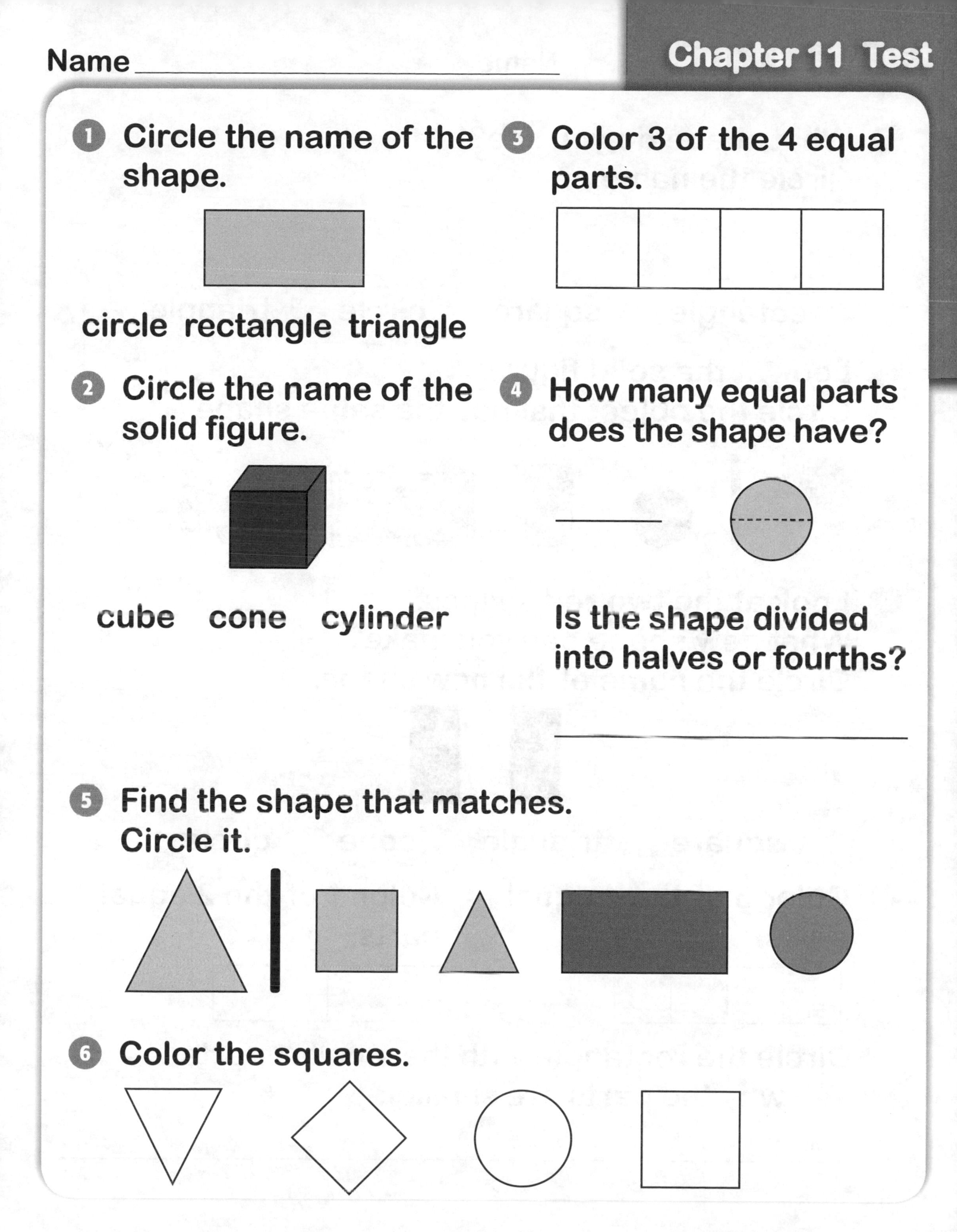

1. Circle the name of the shape.

circle rectangle triangle

2. Circle the name of the solid figure.

cube cone cylinder

3. Color 3 of the 4 equal parts.

4. How many equal parts does the shape have?

______

Is the shape divided into halves or fourths?

____

5. Find the shape that matches. Circle it.

6. Color the squares.

Name ______________________________

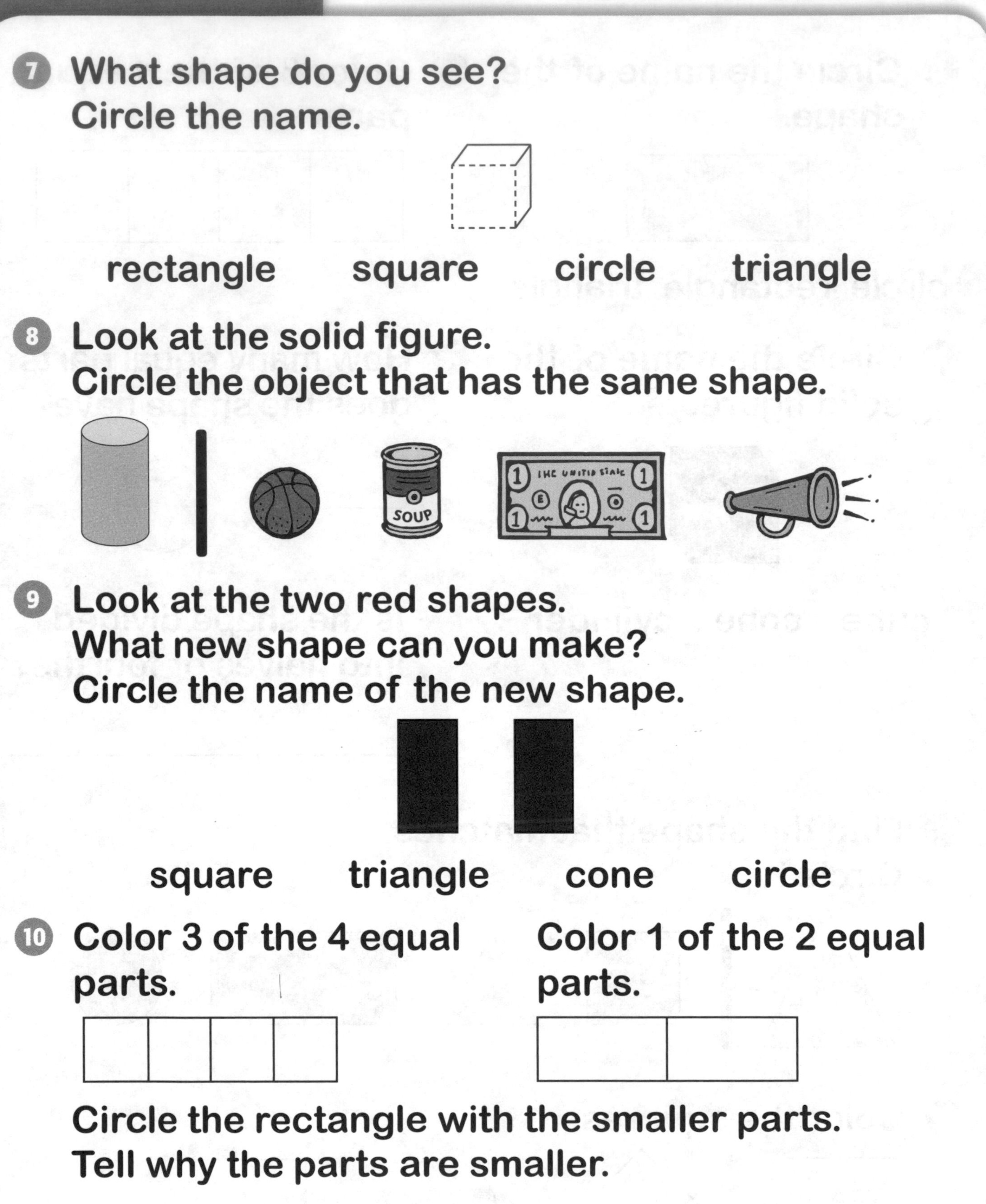

7. What shape do you see?
Circle the name.

rectangle   square   circle   triangle

8. Look at the solid figure.
Circle the object that has the same shape.

9. Look at the two red shapes.
What new shape can you make?
Circle the name of the new shape.

square   triangle   cone   circle

10. Color 3 of the 4 equal parts.

Color 1 of the 2 equal parts.

Circle the rectangle with the smaller parts.
Tell why the parts are smaller.

____

Name ______________________

Data

Do your friends like , , or best?

Sarah asked her family and friends. She made a list.

The list shows data or facts.

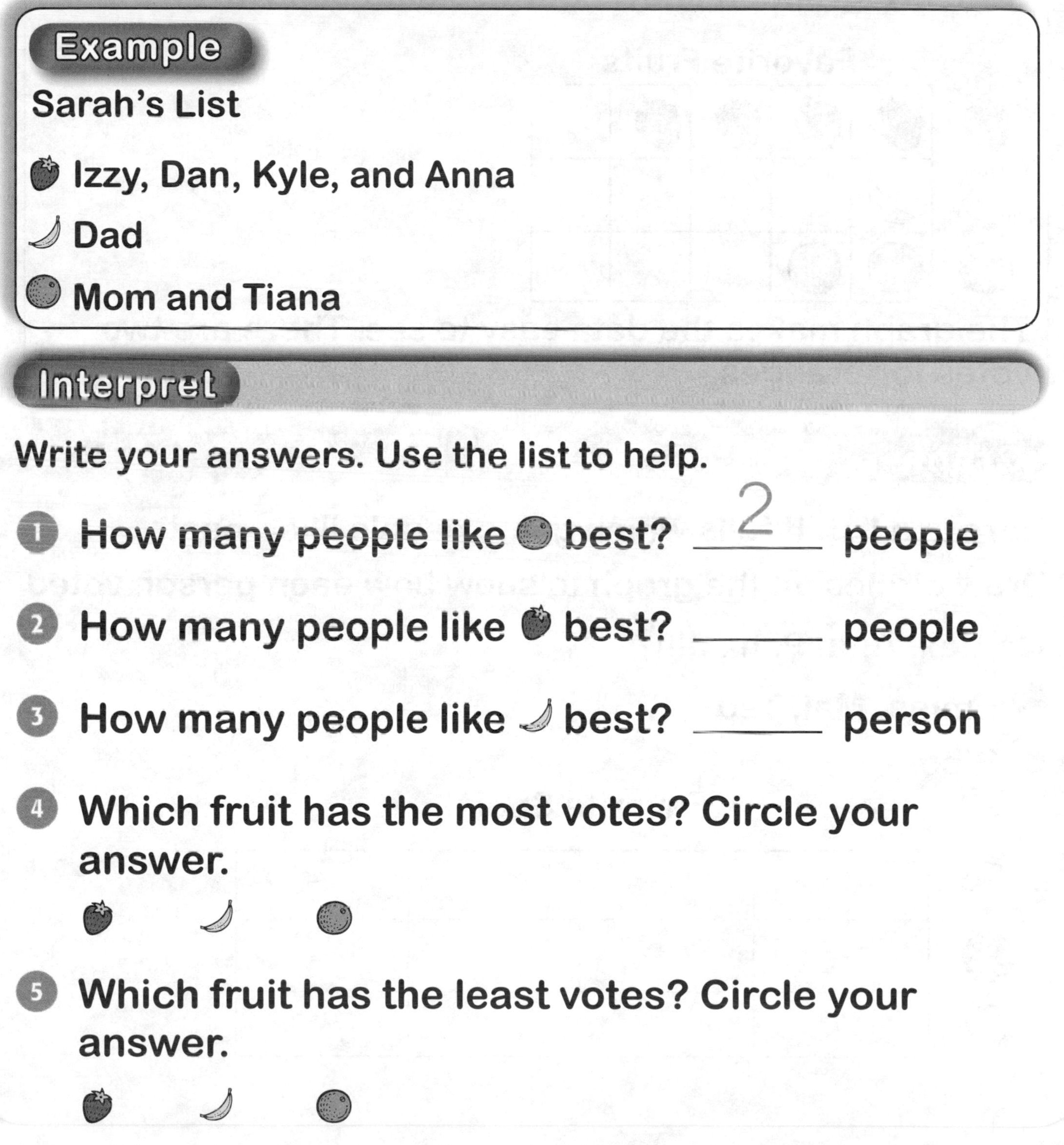

Example

Sarah's List

Izzy, Dan, Kyle, and Anna

Dad

Mom and Tiana

Interpret

Write your answers. Use the list to help.

1. How many people like best? ___2___ people
2. How many people like best? ______ people
3. How many people like best? ______ person
4. Which fruit has the most votes? Circle your answer.
5. Which fruit has the least votes? Circle your answer.

Name ______________________

Organizing and Representing Data

You can show data on a graph.

Example

Here is a graph. It shows Sarah's data on favorite fruits.

Favorite Fruits

The graph makes the data easy to see. There are two votes for oranges.

Graph

Here is a list. It tells which game people liked best.

Draw circles on the graph to show how each person voted.

Alex, Ana, Pam, Jin

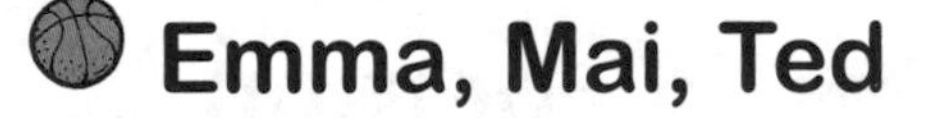

Emma, Mai, Ted

Rico

Favorite Sport

Name ______________________

More Practice with Data

How do children get to school?

Look at the answers 9 children gave.

You can show the answers on a graph.

Graph

Show how children get to school. Look at each child's answer.

Put a ▲ in a row to show each answer.

Kate: I take the bus.	**Josh:** I ride in a car.	**Gamal:** I ride in a car.
Dani: I walk.	**Lina:** I take the bus.	**Abay:** I take the bus.
Emily: I walk.	**Akio:** I take the bus.	**Mike:** I walk.

Ways to Get to School

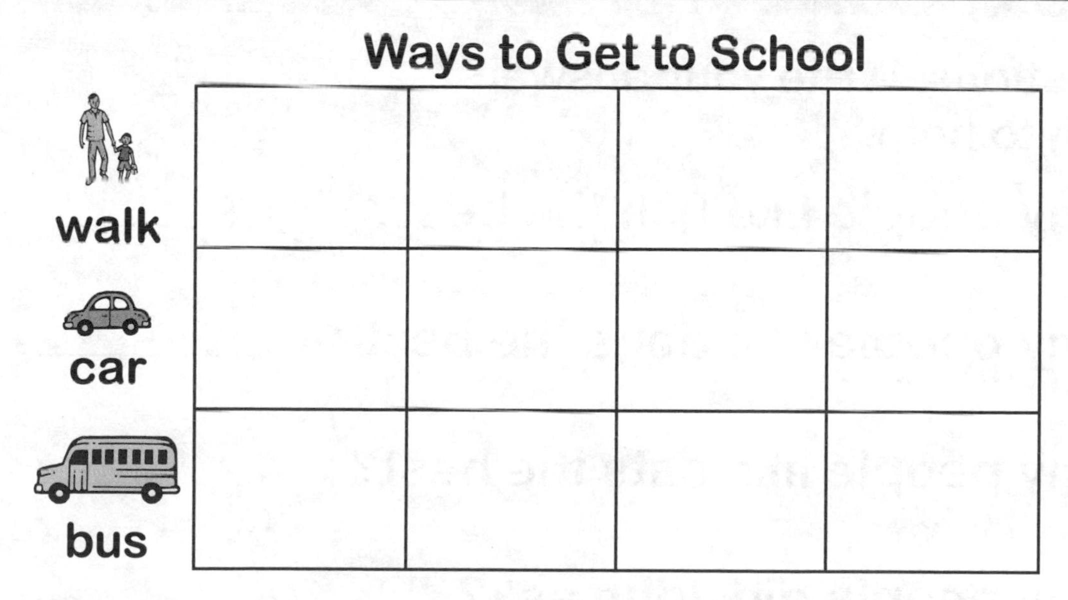

How do most children get to school?

____

Name ______________________

Questions about Data

You can ask and answer questions about data.

Example

John asked some friends about which pet they like best. John shows their answers on this graph.

Favorite Pets

dog	●	●		
cat	●	●	●	
fish	●			

Interpret

Read the questions. Write your answers. Use the graph to help.

1. How many people like fish the best? 1
2. How many people like dogs the best? ______
3. How many people like cats the best? ______
4. How many people did John ask? ______
5. How many people like cats and fish? ______

Name ____________________

Look at the graph. It shows the fruit that children like best.

Read the questions. Write your answers.
Use the graph to help.

1. How many people like (orange) best? ________

2. How many people like (pineapple) best? ________

3. How many people like (apple) best? ________

4. How many people like (apple) and (orange) the best? ________

5. Which fruit has the least votes? Circle your answer.

(apple) (pineapple) (orange)

Name ____________________

Look at the data.

2 people like kite.

3 people like book.

1 person likes ball.

6 Draw ● in the graph to show the data.

Favorite Activity

kite				
book				
ball				

7 Look at the graph. What gets the most votes? Circle your answer.

8 Look at the graph. What is liked less than kite? Circle your answer.

Name ____________________

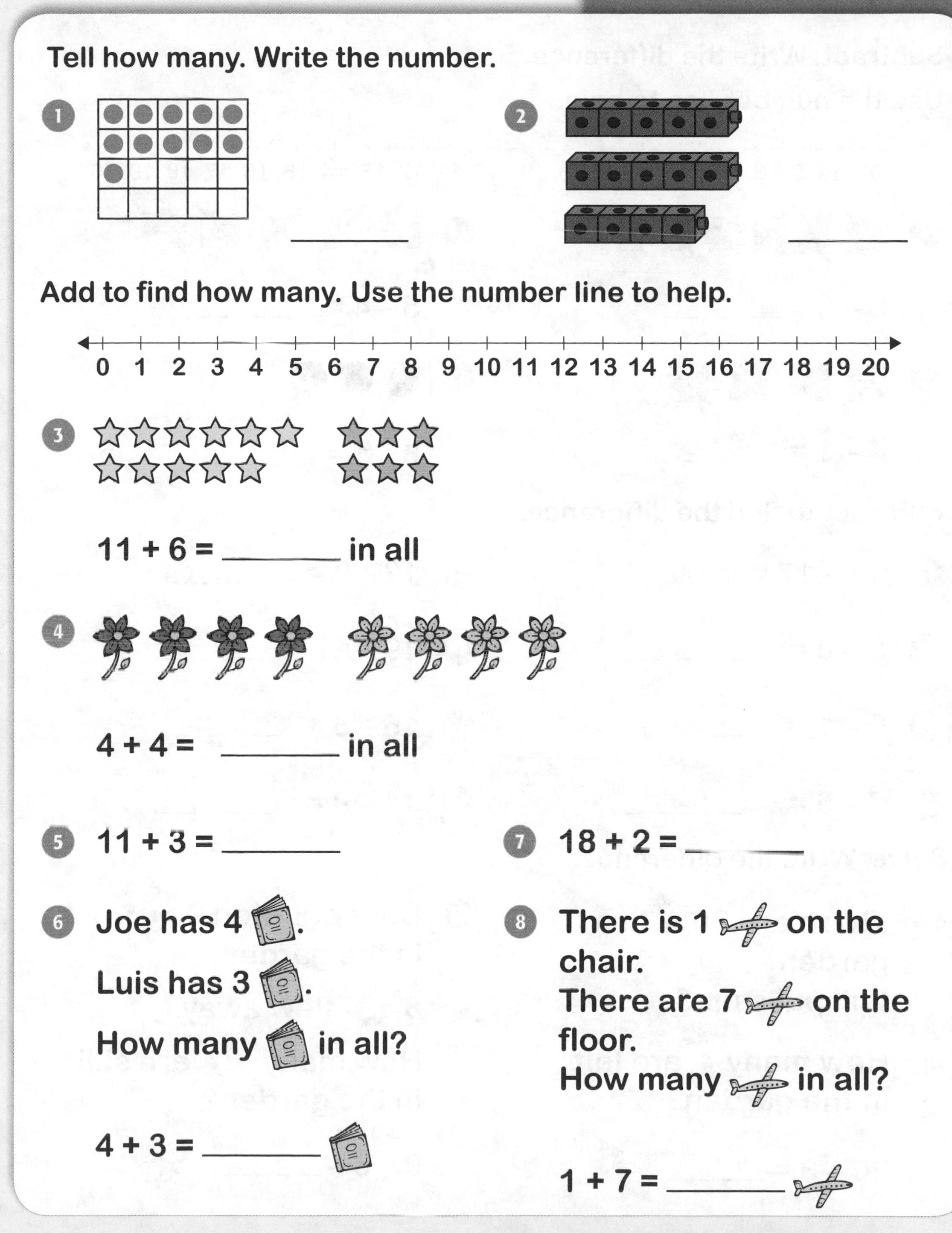

Tell how many. Write the number.

1 ______

2 ______

Add to find how many. Use the number line to help.

0 1 2 3 4 5 6 7 8 9 10 11 12 13 14 15 16 17 18 19 20

3 11 + 6 = ______ in all

4 4 + 4 = ______ in all

5 11 + 3 = ______

6 Joe has 4.
Luis has 3.
How many in all?

4 + 3 = ______

7 18 + 2 = ______

8 There is 1 on the chair.
There are 7 on the floor.
How many in all?

1 + 7 = ______

 Name ______________

Subtract. Write the difference.

Use the number line to help.

0 1 2 3 4 5 6 7 8 9 10 11 12 13 14 15 16 17 18 19 20

9 $6 - 6 =$ ______

11 $5 - 4 =$ ______

10 $4 - 1 =$ ______

12 $3 - 1 =$ ______

Subtract to find the difference.

13 $17 - 17 =$ ______

17 $12 - 9 =$ ______

14 $5 - 0 =$ ______

18 $19 - 17 =$ ______

15 $9 - 1 =$ ______

19 $15 - 8 =$ ______

16 $7 - 6 =$ ______

20 $11 - 9 =$ ______

Solve. Write the difference.

21 There are 20 in a garden.
Will pulls up 3 .

How many are left in the garden?

$20 - 3 =$ ______

22 Cora counted 9 in the garden.

8 flew away.

How many are still in the garden?

$9 - 8 =$ ______

Name ______________________

Count by 1s, 5s, or 10s. Write the missing numbers.

23. 2 3 4 ____ ____ ____ ____ ____ ____

24. 45 50 ____ ____ ____ ____ ____ ____

25. 0 10 20 ____ ____ ____ ____ ____

Look at the chart. Write an addition sentence. Then write a subtraction sentence.

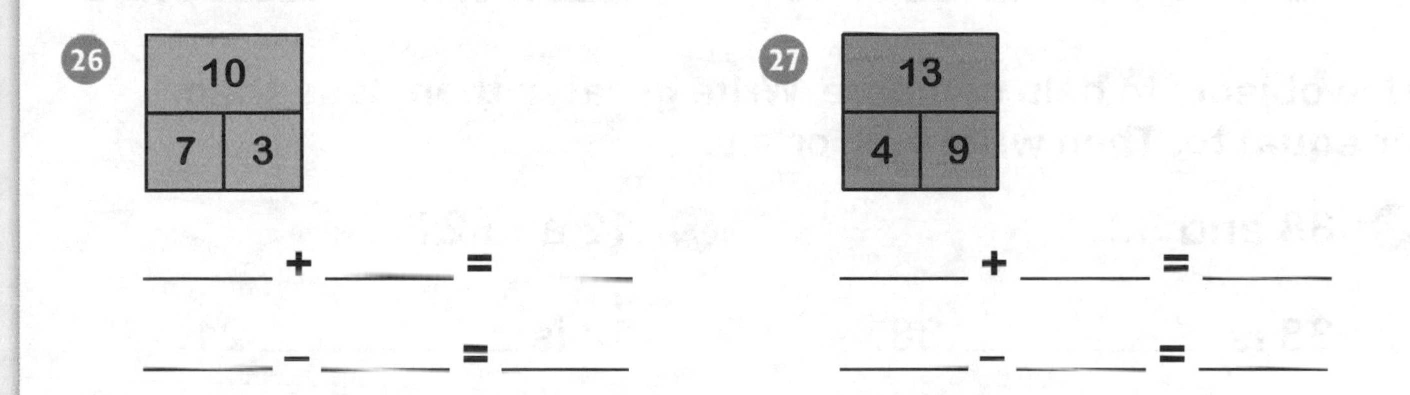

26.

10	
7	3

____ + ____ = ____

____ − ____ = ____

27.

13	
4	9

____ + ____ = ____

____ − ____ = ____

Read the number sentences. Write the missing number.

28. 3 + ? = 14

14 − 3 = ____

3 + ____ = 14

29. 6 + ? = 19

19 − 6 = ____

6 + ____ = 19

Chapters 1-12 Review

Name ___________________________

Count the tens. Tell how many ones. Write your answers.

30

_____ tens = _____ ones

31

_____ tens = _____ ones

Use objects to help compare. Write **greater than, less than,** or **equal to.** Then write >, <, or =

32 38 and 38

38 is ____________ 38.

38 ____________ 38

33 12 and 21

12 is ____________ 21.

12 ____________ 21

Add or subtract. Write the sum or difference.

34
$$\begin{array}{r} 73 \\ +\ \ 4 \\ \hline \end{array}$$

35
$$\begin{array}{r} 80 \\ +11 \\ \hline \end{array}$$

36
$$\begin{array}{r} 50 \\ -30 \\ \hline \end{array}$$

Find 10 more or 10 less. Write your answer.

37 76

10 more than 76 is _______.

10 less than 76 is _______.

Name ______________________________

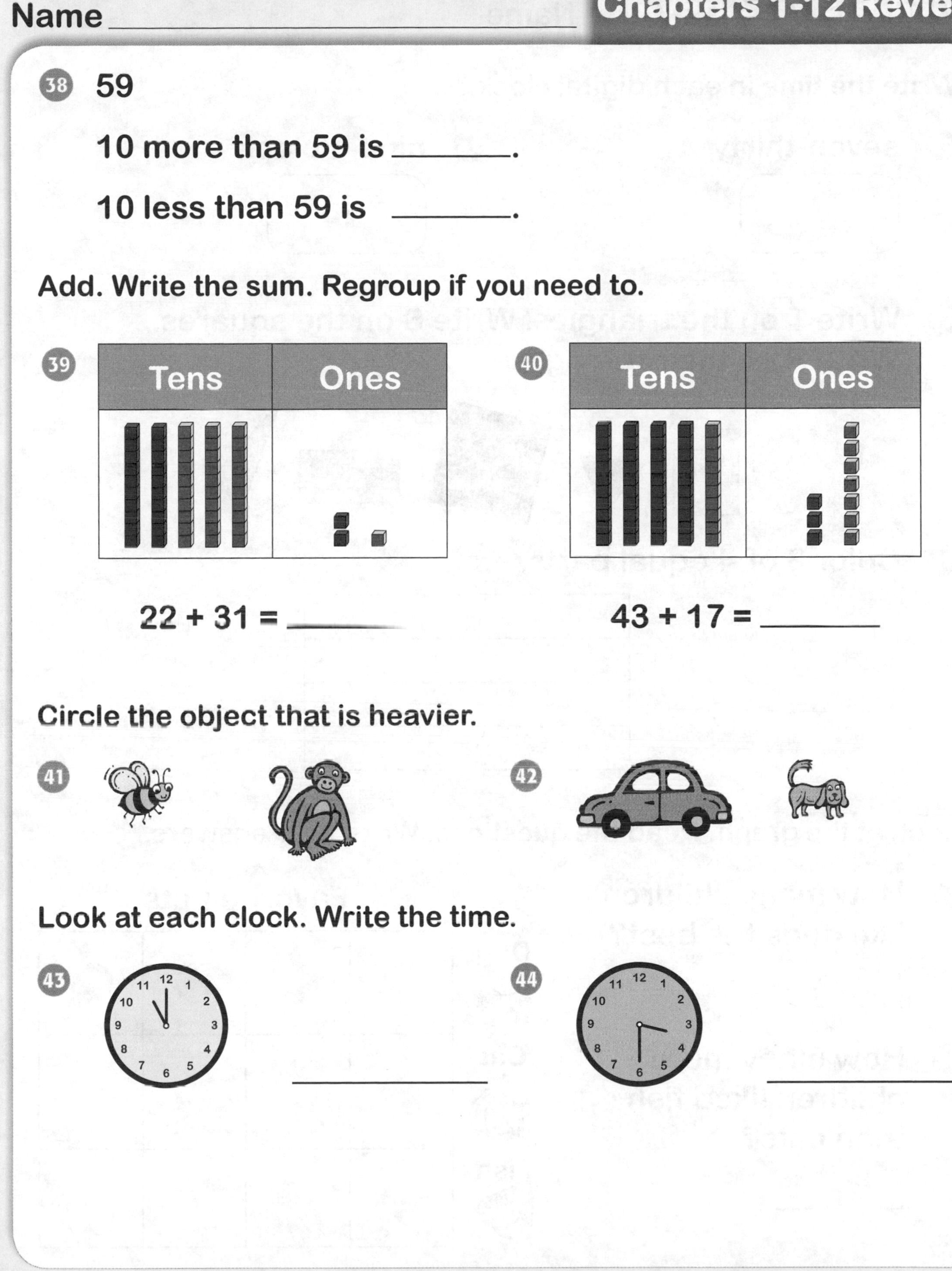

38 59

10 more than 59 is ______.

10 less than 59 is ______.

Add. Write the sum. Regroup if you need to.

39

Tens	Ones

$22 + 31 =$ ______

40

Tens	Ones

$43 + 17 =$ ______

Circle the object that is heavier.

41

42

Look at each clock. Write the time.

43 ______________

44 ______________

 Name ____________________

Write the time in each digital clock.

45 seven-thirty

46 nine o'clock

47 Write T on the triangles. Write S on the squares. Write R on the rectangle.

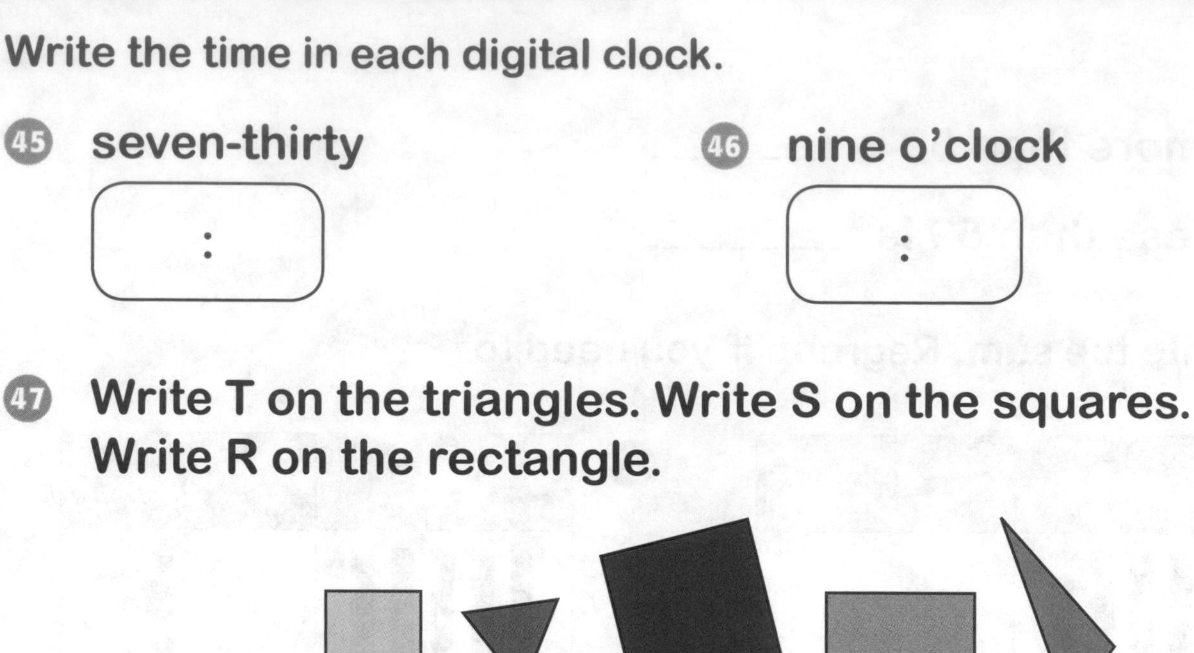

48 Color 3 of 4 equal parts.

Look at the graph. Read the questions. Write your answers.

49 How many children like dogs the best?

____

50 How many more children liked fish than cats?

____

Favorite Pets

Dog	●	●	●	●
Cat	●			
Fish	●	●		

Add: To put groups together and tell how many in all. *(p. 16)*

●● + ○○○ = ●●○○○

2 + 3 = 5

Addends: Numbers you add. *(p. 58)*

2 + 3 = 5
↑ ↑
Addends

Circle: ● *(p. 92)*

Compare Lengths: To measure objects and tell which is longer. *(p. 86)*

Compare Numbers: To tell if a number is greater than (>), less than (<), or equal to (=) another number. *(p. 70)*

Compare Shapes: To tell how shapes are alike or different. *(p. 96)*

Cone: *(p. 100)*

Cube: *(p. 100)*

Cylinder: *(p. 100)*

Data: Information you collect. *(p. 107)*

Difference: The answer to a subtraction problem. *(p. 24)*

Digital Clock: A clock that shows time with only digits. *(p. 89)*

Equal: To have the same value or amount. *(p. 64)*

Equal Parts: Parts that are the same number and size. *(p. 103)*

The square has 4 equal parts.

Equal Sign (=): A symbol that shows when numbers are equal. *(p. 64)*

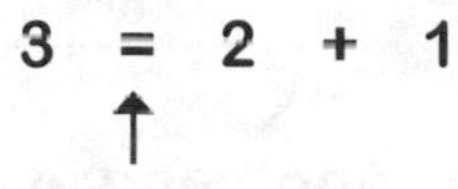

Fourth, Fourths: Four equal parts of a whole. *(p. 103)*

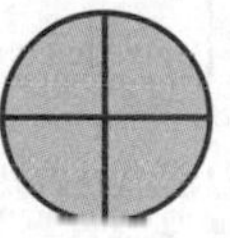

The circle is divided into fourths.

Graph: A way to show data. *(p. 108)*

The farmer has 4 horses and 2 pigs.

Greater Than (>): 3 > 2

3 is greater than 2. *(p. 70)*

Half, Halves: Two equal parts of a whole. *(p. 103)*

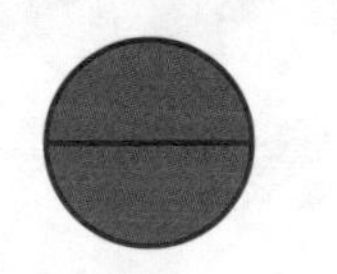

The circles are divided in half.

Heavy, Heavier:

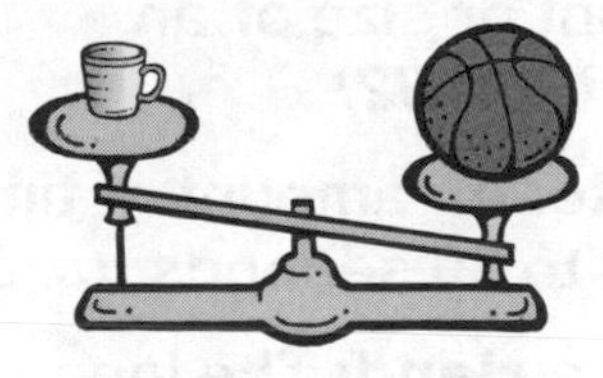

The ball is heavier than the mug. *(p. 82)*

Hour: 60 minutes. *(p. 88)*

Hour Hand: The short hand on a clock. *(p. 88)*

Hour Hand

Inch: A unit used to measure length. *(p. 82)*

Length: How long something is. *(p. 82)*

Less Than (<): 3 < 4

3 is less than 4. *(p. 70)*

Light, Lighter:

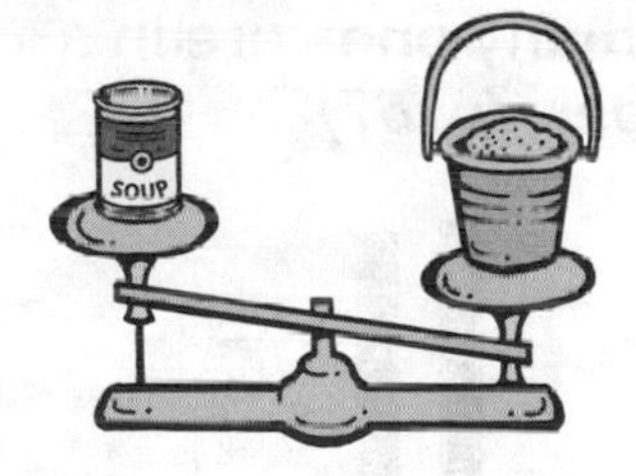

The can is lighter than the pail. *(p. 82)*

Longest:

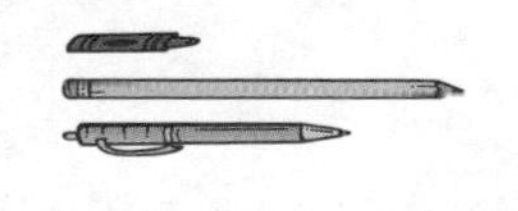

The pencil is longest. *(p. 85)*

Measure: To find the amount or size of an object. *(p. 82)*

Minute: An amount of time equal to 60 seconds. *(p. 88)*

Minute Hand: The long hand on a clock. *(p. 88)*

Minute → Hand

Number Sentence: A way to show sums or differences. *(p. 16)*

3 + 4 = 7

8 – 6 = 2

O'Clock: What to say when a minute hand points to 12. *(p. 88)*

It is 4 o'clock.

Ones: A digit that shows how many ones are in a number. *(p. 67)*

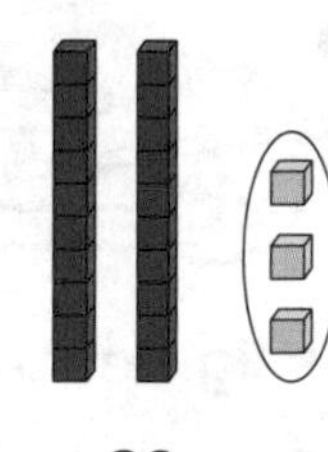

23

There are 3 ones.

Place Value Chart: A chart that shows how many tens and ones in a number. *(p. 75)*

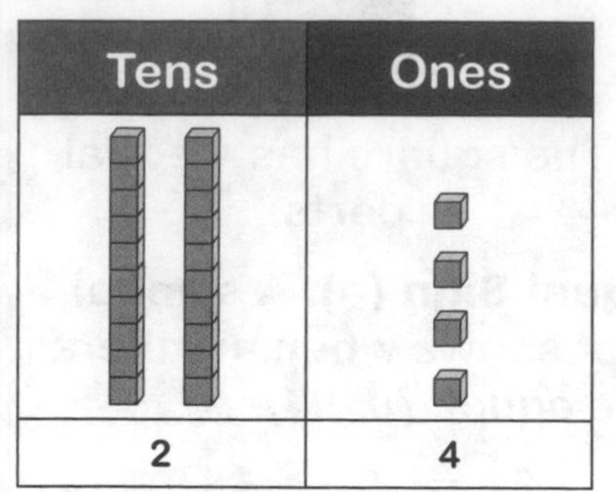

Tens	Ones
2	4

Quarter: Another way to say "fourth." *(p. 103)*

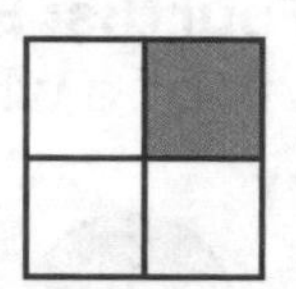

One quarter is green.

Rectangle: *(p. 92)*

Rectangular Prism: *(p. 100)*

Ruler: A tool used to measure length. *(p. 82)*

Scale: A tool used to measure weight. *(p. 82)*

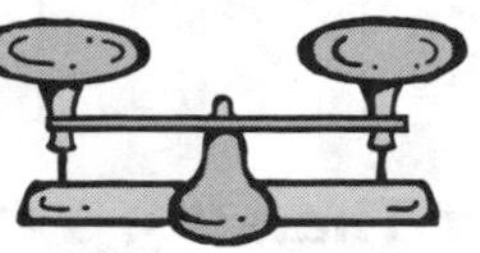

Shape: 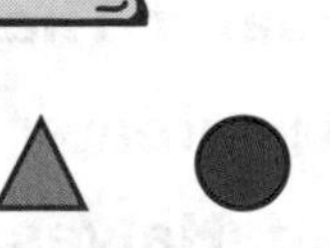 *(p. 92)*

Shortest:

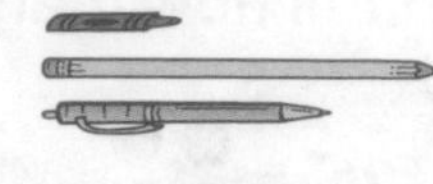

The red crayon is shortest. *(p. 85)*

Solid Figure: A shape that is not flat.

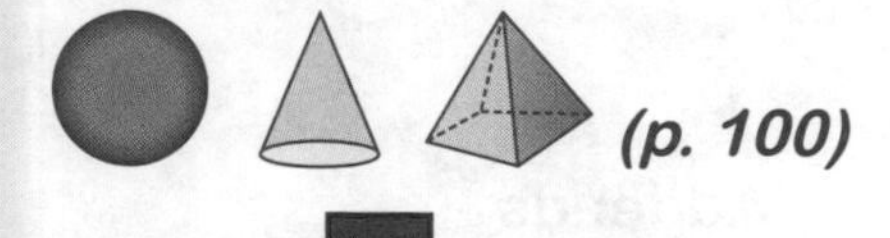

(p. 100)

Square: *(p. 92)*

Sort: To put matching objects into groups. *(p. 87)*

Subtract: To take groups away and tell how many are left. *(p. 24)*

4 – 2 = 2 left over

Sum: The answer to an addition problem. *(p. 16)*

Tens: A digit that shows how many tens are in a number. *(p. 67)*

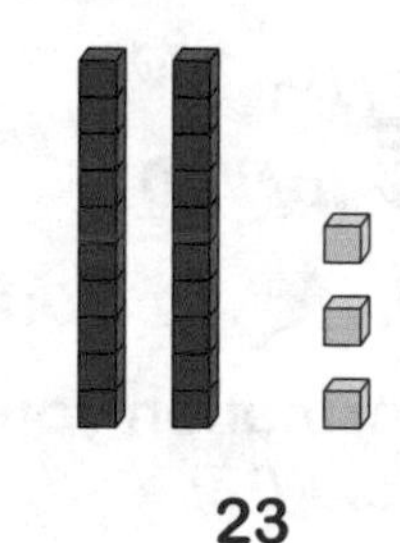

23

There are 2 tens.

Triangle: *(p. 92)*

Weight: How heavy something is. *(p. 82)*

Chapter 1

Chapter 1 • Lesson 1
Page 8
1. circle 3
2. circle 4
3. 5
4. 2
5. 1
6. 0
7. 4
8. 3

Chapter 1 • Lesson 2
Page 9
1. circle 7
2. circle 10
3. 8
4. 6
5. 7
6. 10
7. 9
8. 6

Chapter 1 • Lesson 3
Page 10
1. circle 12
2. circle 15
3. 12
4. 14
5. 11
6. 13
7. 15
8. 12

Chapter 1 • Lesson 4
Page 11
1. circle 16
2. circle 19
3. 17
4. 18
5. 20
6. 17
7. 16
8. 18

Chapter 1 • Lesson 5
Page 12
1. 3
2. 6
3. 16
4. 13
5. 1
6. 20
7. 14
8. 9

Chapter 1 • Lesson 6
Page 13
1. 7
2. 12
3. 4
4. 19
5. 0
6. 10

Chapter 1 Test
Pages 14-15
1. 4
2. 8
3. 5
4. 15
5. 11
6. 7
7. 2
8. 0
9. 14
10. 19
11. 17
12. 10
13. 9
14. 6
15. 13
16. 20
17. 3
18. 1
19. 18
20. 12

Chapter 2

Chapter 2 • Lesson 1
Page 16
1. 4
2. 5
3. 6
4. 3
5. 1
6. 6

Chapter 2 • Lesson 2
Page 17
1. 10
2. 9
3. 11
4. 12
5. 7
6. 8

Chapter 2 • Lesson 3
Page 18
1. 5
2. 9
3. 1
4. 12
5. 11
6. 7
7. 6
8. 3

Chapter 2 • Lesson 4
Page 19
1. 15
2. 19
3. 11
4. 13
5. 20
6. 12

Chapter 2 • Lesson 5
Page 20
1. 6
2. 11
3. 15
4. 2
5. 20
6. 14

Chapter 2 • Lesson 6
Page 21
1. 9
2. 6
3. 14
4. 10

Chapter 2 Test
Pages 22-23
1. 4
2. 3
3. 5
4. 11
5. 8
6. 2
7. 5
8. 12
9. 10
10. 9
11. 2
12. 9
13. 12
14. 8
15. 19
16. 13
17. 17
18. 14
19. 7
20. 12

Chapter 3

Chapter 3 • Lesson 1
Page 24
1. 2
2. 1
3. 2
4. 3
5. 3
6. 3

Chapter 3 • Lesson 2
Page 25
1. 7
2. 6
3. 2
4. 11

Chapter 3 • Lesson 2 (continued)

5. 5
6. 3
7. 5
8. 1
9. 6
10. 4

Chapter 3 • Lesson 3
Page 26

1. 1
2. 8
3. 0
4. 3
5. 6
6. 5
7. 2
8. 11
9. 0
10. 4

Chapter 3 • Lesson 4
Page 27

1. 10
2. 9
3. 6
4. 1
5. 4
6. 2
7. 11
8. 16
9. 12
10. 7

Chapter 3 • Lesson 5
Page 28

1. 3
2. 10
3. 13
4. 7
5. 2
6. 9
7. 12
8. 5
9. 1
10. 11
11. 0
12. 8

Chapter 3 • Lesson 6
Page 29

1. 5
2. 2
3. 20, 21, 22
4. 45, 46, 47
5. 19, 20, 21
6. 25, 26, 27, 28
7. 47, 48, 49, 50

Chapter 3 Test
Pages 30-31

1. 1
2. 3
3. 2
4. 4
5. 3
6. 6
7. 8
8. 7
9. 1
10. 0
11. 5
12. 4
13. 11
14. 10
15. 8
16. 3
17. 0
18. 6
19. 5
20. 14
21. 16
22. 13
23. 8
24. 5
25. 3
26. 5

Chapter 4

Chapter 4 • Lesson 1
Page 32

1. 2, 3, 4
2. 6, 7, 8
3. 8, 9, 10
4. 16, 17, 18
5. 13, 14, 15
6. 10, 11, 12
7. 18, 19, 20
8. 11, 12, 13

Chapter 4 • Lesson 2
Page 33

0

1	2	3	4	5	6	7	8	9	10
11	12	13	14	15	16	17	18	19	20
21	22	23	24	25	26	27	28	29	30
31	32	33	34	35	36	37	38	39	40
41	42	43	44	45	46	47	48	49	50

Chapter 4 • Lesson 3
Page 34

1. 2, 3, 4, 5, 6
2. 9, 10, 11, 12
3. 16, 17, 18, 19
4. 23, 24, 25, 26
5. 30, 31, 32, 33
6. 36, 37, 38, 39
7. 42, 43, 44, 45
8. 47, 48, 49, 50

Chapter 4 • Lesson 4
Page 35

1. 8, 9, 10
2. 24, 25, 26
3. 2, 3, 4
4. 2, 3, 4, 5
5. 33, 34, 35, 36
6. 20, 21, 22
7. 45, 46, 47
8. 19, 20, 21
9. 25, 26, 27, 28
10. 47, 48, 49, 50

Chapter 4 Test
Pages 36-37

1. 1
2. 6
3. 10
4. 4
5. 16
6. 19
7. 8
8. 12
9.

0

1	2	3	4	5	6	7	8	9	10
11	12	13	14	15	16	17	18	19	20
21	22	23	24	25	26	27	28	29	30
31	32	33	34	35	36	37	38	39	40
41	42	43	44	45	46	47	48	49	50

10. 2, 3, 4, 5
11. 15, 16, 17, 18
12. 36, 37, 38, 39
13. 29, 30, 31, 32
14. 41, 42, 43, 44
15. 20, 21, 22, 23
16. 8, 9, 10, 11
17. 47, 48, 49, 50
18. 7, 8, 9, 10
19. 19, 20, 21, 22
20. 30, 31, 32, 33
21. 21, 22, 23, 24
22. 42, 43, 44, 45
23. 16, 17, 18, 19
24. 47, 48, 49, 50
25. 3, 4, 5, 6

Chapter 5

Chapter 5 • Lesson 1
Page 38

1. 51, 52, 53
2. 55, 56, 57
3. 62, 63, 64
4. 71, 72, 73
5. 54, 55, 56
6. 59, 60, 61
7. 67, 68, 69
8. 73, 74, 75

Chapter 5 • Lesson 2
Page 39

1. 77, 78, 79
2. 83, 84, 85

3. 98, 99, 100
4. 79, 80, 81
5. 91, 92, 93
6. 78, 79, 80
7. 89, 90, 91
8. 95, 96, 97

Chapter 5 • Lesson 3
Page 40

1. 10, 20
2. 70, 80
3. 30, 40
4. 90, 100
5. 50, 60, 70
6. 80, 90, 100
7. 50, 70, 80
8. 30, 40, 60

Chapter 5 • Lesson 4
Page 41

1	2	3	4	5	6	7	8	9	10
11	12	13	14	15	16	17	18	19	20
21	22	23	24	25	26	27	28	29	30
31	32	33	34	35	36	37	38	39	40
41	42	43	44	45	46	47	48	49	50
51	52	53	54	55	56	57	58	59	60
61	62	63	64	65	66	67	68	69	70
71	72	73	74	75	76	77	78	79	80
81	82	83	84	85	86	87	88	89	90
91	92	93	94	95	96	97	98	99	100

Chapter 5 • Lesson 5
Page 42

1. 4, 5, 6, 9
2. 78, 80, 81, 83
3. 93, 95, 98, 99
4. 51, 53, 55, 57
5. 83, 85, 86, 88
6. 70, 71, 72, 73
7. 19, 20, 23, 25
8. 34, 35, 37, 40

Chapter 5 Test
Pages 43-44

1.

1	2	3	4	5	6	7	8	9	10
11	12	13	14	15	16	17	18	19	20
21	22	23	24	25	26	27	28	29	30
31	32	33	34	35	36	37	38	39	40
41	42	43	44	45	46	47	48	49	50
51	52	53	54	55	56	57	58	59	60
61	62	63	64	65	66	67	68	69	70
71	72	73	74	75	76	77	78	79	80
81	82	83	84	85	86	87	88	89	90
91	92	93	94	95	96	97	98	99	100

2. 55, 56, 57
3. 62, 63, 64
4. 58, 59, 60
5. 71, 72, 73
6. 73, 74, 75
7. 81, 82, 83
8. 98, 99, 100
9. 79, 80, 81
10. 89, 90, 91
11. 91, 92, 93
12. 30, 40, 50
13. 70, 80, 90
14. 50, 60, 70
15. 80, 90, 100
16. 51, 52
17. 86, 87, 88
18. 42, 44
19. 41, 42
20. 98, 99
21. 35, 37
22. 69, 70
23. 81, 83
24. 92, 93, 94
25. 18, 19

Chapter 6

Chapter 6 • Lesson 1
Page 45

1. 101, 102, 103
2. 106, 107, 108
3. 111, 112, 113
4. 112, 113, 114
5. 105, 106, 107
6. 102, 103, 104

Chapter 6 • Lesson 2
Page 46

1. 30, 40
2. 100, 110
3. 20, 30, 40, 50
4. 80, 90, 100, 110
5. 0, 10, 20, 30
6. 50, 60, 70, 80

Chapter 6 • Lesson 3
Page 47

1. 15, 20
2. 40, 45
3. 65, 70
4. 35, 40, 45, 50
5. 60, 65, 70, 75
6. 80, 85, 90, 95

Chapter 6 • Lesson 4
Page 48

1. 40, 41, 42
2. 95, 100, 105
3. 108, 109, 110
4. 35, 40, 45
5. 60, 70, 80
6. 100, 110, 120

Chapter 6 • Lesson 5
Page 49

0	5	10	15	20	25
30	35	40	45	50	55
60	65	70	75	80	85
90	95	100	105	110	115
120					

0	10	20	30	40	50
60	70	80	90	100	110
120					

Chapter 6 Test
Pages 50-51

1. 105, 106, 107
2. 108, 109, 110
3. 116, 117, 118
4. 43, 44, 45
5. 58, 59, 60
6. 69, 70, 71
7. 99, 100, 101
8. 77, 78, 79
9. 20, 30, 40
10. 100, 110, 120
11. 70, 80, 90
12. 60, 70, 80
13. 90, 100, 110
14. 40, 50, 60
15. 50, 60, 70
16. 80, 90, 100
17. 35, 40, 45
18. 20, 25, 30
19. 75, 80, 85
20. 10, 15, 20
21. 90, 95, 100
22. 45, 50, 55
23. 110, 115
24. 25, 30, 35
25.

0	5	10	15	20	25
30	35	40	45	50	55
60	65	70	75	80	85
90	95	100	105	110	115
120					

26.

0	10	20	30	40	50
60	70	80	90	100	110
120					

Chapters 1-6 Review
Pages 52-55

1. 3
2. 5
3. 7
4. 15
5. 20
6. 6
7. 4
8. 2
9. 16
10. 6
11. 5
12. 6
13. 6
14. 2
15. 3
16. 8
17. 13
18. 10
19. 11
20. 12
21. 20
22. 20
23. 5
24. 1
25. 2
26. 1
27. 1
28. 13
29. 7
30. 0
31. 8
32. 7
33. 2
34. 17
35. 8, 9, 10, 11
36. 75, 80, 90, 95
37. 30, 50, 60, 70
38. 35, 40, 45, 50
39. 9
40. 6
41. 16
42. 14

Chapter 7

Chapter 7 • Lesson 1
Page 56

1. 4, 4
2. 7, 7
3. 5
4. 6
5. 18
6. 19

Chapter 7 • Lesson 2
Page 57

1. 7, 7
2. 9, 9
3. 8, 8
4. 13, 13

Chapter 7 • Lesson 3
Page 58

1. 1, 1
2. 2, 2
3. 9, 9
4. 3, 3

Chapter 7 • Lesson 4
Page 59

1. 9
2. 11
3. 18
4. 6
5. 18
6. 6
7. 14
8. 5

Chapter 7 • Lesson 5
Page 60

1. draw 5 circles 12, 12
2. draw 7 circles 16, 16

Chapter 7 • Lesson 6
Page 61

1. 6 + 1 = 7 and 7 – 1 = 6
2. 8 + 9 = 17 or 9 + 8 = 17
17 – 9 = 8 or 17 – 8 = 9

Chapter 7 • Lesson 7
Page 62

Possible answers:
1. 10 + 5 = 15
2. 4 + 4, 1 + 7, 2 + 6, 0 + 8
3. 10 + 3, 12 + 1, 11 + 2,
8 + 5, 7 + 6, 13 + 0
4. 9 + 2, 10 + 1, 8 + 3,
7 + 4, 11 + 0
5. 4 + 6, 1 + 9, 2 + 8, 5 + 5, 10 + 0
6. 3 + 2, 1 + 4
7. 3 + 3, 1 + 5, 6 + 0
8. 7 + 2, 6 + 3, 5 + 4, 9 + 0

Chapter 7 • Lesson 8
Page 63

1. 9
2. 6

Chapter 7 • Lesson 9
Page 64

1. 6
2. 23
3. 103
4. 82
5. not =
6. =
7. ■
8. =

Chapter 7 Test
Pages 65-66

1. 12, 12
2. 18, 18
3. 8
4. 4
5. 8
6. 13
7. 4, 4
8. 13, 13
9. Possible answers:
5 + 6, 4 + 7, 3 + 8,
1 + 10, 11 + 0
10. Possible answers:
1 + 6, 2 + 5, 7 + 0
11. =
12. =
13. not =
14. =
15. draw 5 circles 13, 13
16. draw 6 circles 15, 15
17. 11
18. 6
19. 2 + 4 = 6 or 4 + 2 = 6
20. 6 – 4 = 2 or 6 – 2 = 4
21. 8
22. 7

Chapter 8

Chapter 8 • Lesson 1
Page 67

1. 30 ones, 3 tens
2. 50 ones, 5 tens

Chapter 8 • Lesson 2
Page 68

1. 1 ten, 5 ones
2. 1 ten, 4 ones
3. 1 ten, 8 ones
4. 1 ten, 2 ones
5. 1 ten, 6 ones
6. 1 ten, 9 ones
7. 1 ten, 7 ones
8. 1 ten, 3 ones

Chapter 8 • Lesson 3
Page 69

1. 5 tens = 50 ones
2. 2 tens = 20 ones
3. 7 tens = 70 ones
4. 4 tens = 40 ones

Chapter 8 • Lesson 4
Page 70
1. less than, <
2. greater than, >

Chapter 8 Test
Pages 71-72
1. 40 ones, 4 tens
2. 20 ones, 2 tens
3. 1 ten, 3 ones
4. 1 ten, 7 ones
5. 1 ten, 1 one
6. 1 ten, 9 ones
7. 4 tens, 40 ones
8. 6 tens, 60 ones
9. less than, <
10. greater than, >
11. less than, <
12. equal to, =

Chapter 9

Chapter 9 • Lesson 1
Page 73
1. 16
2. 37
3. 23
4. 15
5. 48
6. 17

Chapter 9 • Lesson 2
Page 74
1. 28
2. 44
3. 17
4. 42

Chapter 9 • Lesson 3
Page 75
1. 2, 6, 26
2. 5, 2, 52

Chapter 9 • Lesson 4
Page 76
1. 34
2. 56

Chapter 9 • Lesson 5
Page 77
1. 52, 32
2. 39, 19
3. 46, 26

Chapter 9 • Lesson 6
Page 78
1. 42
2. 25
3. 54
4. 43
5. 62
6. 33

Chapter 9 • Lesson 7
Page 79
1. 30
2. 10
3. 30
4. 20
5. 50
6. 70

Chapter 9 Test
Pages 80-81
1. 15
2. 36
3. 20, 7, 27
4. 40, 1, 41
5. 57
6. 43
7. 34, 14
8. 91, 71
9. 39
10. 57
11. 83
12. 6
13. 10
14. 20
15. 70
16. 48
17. 68

Chapter 10

Chapter 10 • Lesson 1
Page 82
1. Answers will vary.
2. circle the dog
3. circle the apple

Chapter 10 • Lesson 2
Page 83
1. Answers will vary.
2. circle the pineapple

Chapter 10 • Lesson 3
Page 84
1. circle the pot
2. circle the bucket
3. circle the larger bowl
4. circle the bag

Chapter 10 • Lesson 4
Page 85
1. Color last crayon in the row blue.
2. Color first crayon in the row green.
3. Color first pencil in the row red.
4. Color last pencil in the row blue. Color middle pencil orange.

Chapter 10 • Lesson 5
Page 86
1. circle the top pen
2. circle the top paperclip

Chapter 10 • Lesson 6
Page 87
1. Group 1: 3 yellow hearts, 1 yellow circle; 4 shapes
2. Group 2: 2 purple triangles, 2 purple circles; 4 shapes

Chapter 10 • Lesson 7
Page 88
1. 3:30
2. 5:00
3. 2:00
4. 11:30

Chapter 10 • Lesson 8
Page 89
1. 2:00
2. 1:30
3. 8:00
4. 7:00
5. 5:30
6. 3:00

Chapter 10 Test
Pages 90-91
1. Answers will vary.
2. Answers will vary.
3. Answers will vary.
4. circle the spoon
5. circle the pot
6. Color last crayon in the row red.
7. Color first crayon in the row yellow.
8. circle the glass
9. 10:30
10. 9:30
11. Group 1: 1 purple triangle, 2 purple circles, 3 shapes; Group 2: 1 yellow circle, 1 yellow square, 2 shapes

Chapter 11

Chapter 11 • Lesson 1
Page 92
1. circle word triangle
2. circle word circle
3. circle the yellow rectangle
4. color the first shape, the circle, green

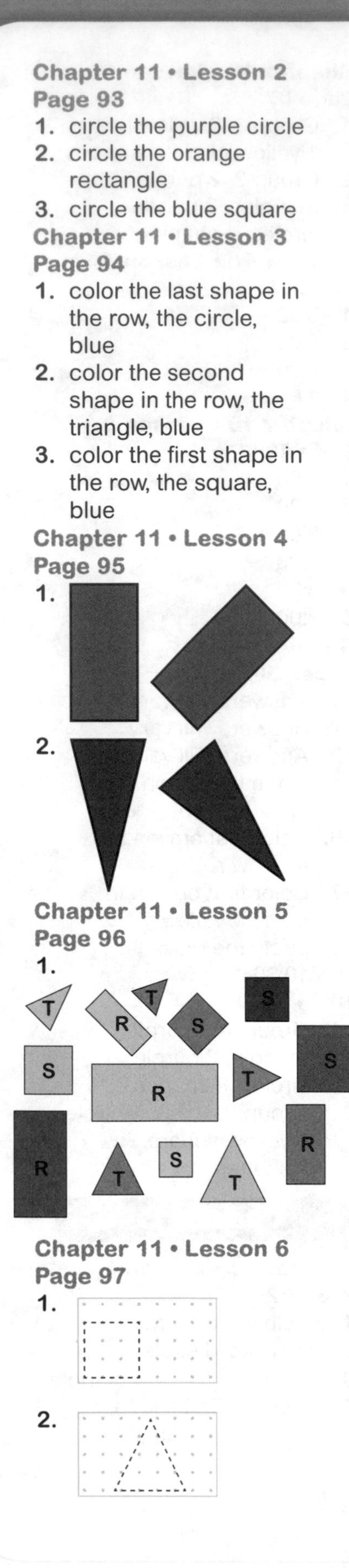

Chapter 11 • Lesson 2
Page 93

1. circle the purple circle
2. circle the orange rectangle
3. circle the blue square

Chapter 11 • Lesson 3
Page 94

1. color the last shape in the row, the circle, blue
2. color the second shape in the row, the triangle, blue
3. color the first shape in the row, the square, blue

Chapter 11 • Lesson 4
Page 95

1.
2.

Chapter 11 • Lesson 5
Page 96

1.

Chapter 11 • Lesson 6
Page 97

1.
2.

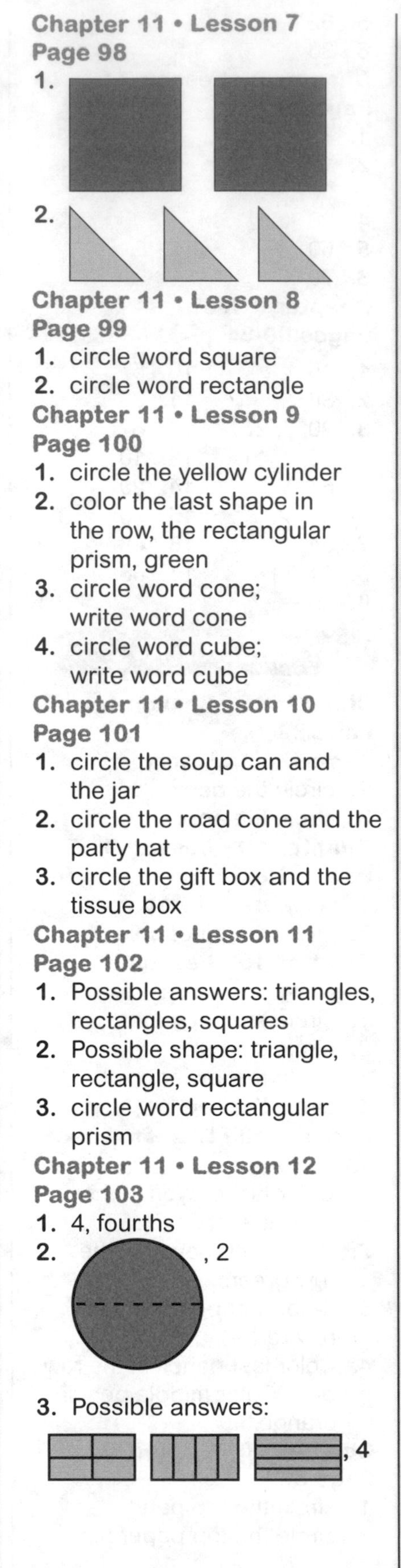

Chapter 11 • Lesson 7
Page 98

1.
2.

Chapter 11 • Lesson 8
Page 99

1. circle word square
2. circle word rectangle

Chapter 11 • Lesson 9
Page 100

1. circle the yellow cylinder
2. color the last shape in the row, the rectangular prism, green
3. circle word cone; write word cone
4. circle word cube; write word cube

Chapter 11 • Lesson 10
Page 101

1. circle the soup can and the jar
2. circle the road cone and the party hat
3. circle the gift box and the tissue box

Chapter 11 • Lesson 11
Page 102

1. Possible answers: triangles, rectangles, squares
2. Possible shape: triangle, rectangle, square
3. circle word rectangular prism

Chapter 11 • Lesson 12
Page 103

1. 4, fourths
2. , 2
3. Possible answers: , 4

Chapter 11 • Lesson 13
Page 104

1.
2.
3.
4.
5. The square in question 4 has smaller parts. It is the same size as the other one, but it has more parts.

Chapter 11 Test
Pages 105-106

1. circle word rectangle
2. circle word cube
3.
4. 2, halves
5. circle the second shape, the blue triangle
6. color the second and the fourth shapes, the squares
7. circle word square
8. circle the soup can
9. circle word square
10.

circle the shape with 4 parts

It is the same size as the other one, but it has more parts.

Chapter 12

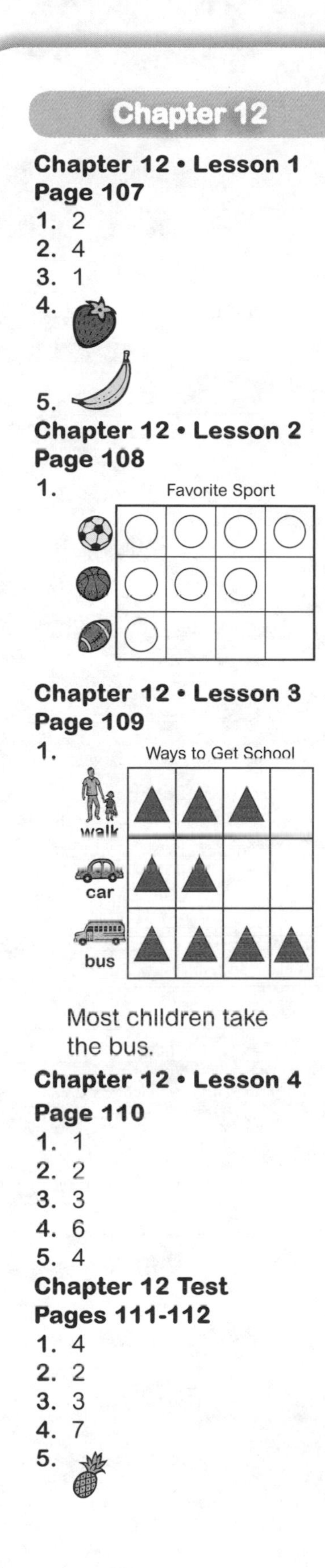

Chapter 12 • Lesson 1
Page 107

1. 2
2. 4
3. 1
4. (strawberry)
5. (banana)

Chapter 12 • Lesson 2
Page 108

1. Favorite Sport

soccer	○	○	○	○
basketball	○	○	○	
football	○			

Chapter 12 • Lesson 3
Page 109

1. Ways to Get School

walk	▲	▲	▲	
car	▲	▲		
bus	▲	▲	▲	▲

Most children take the bus.

Chapter 12 • Lesson 4
Page 110

1. 1
2. 2
3. 3
4. 6
5. 4

Chapter 12 Test
Pages 111-112

1. 4
2. 2
3. 3
4. 7
5. (pineapple)

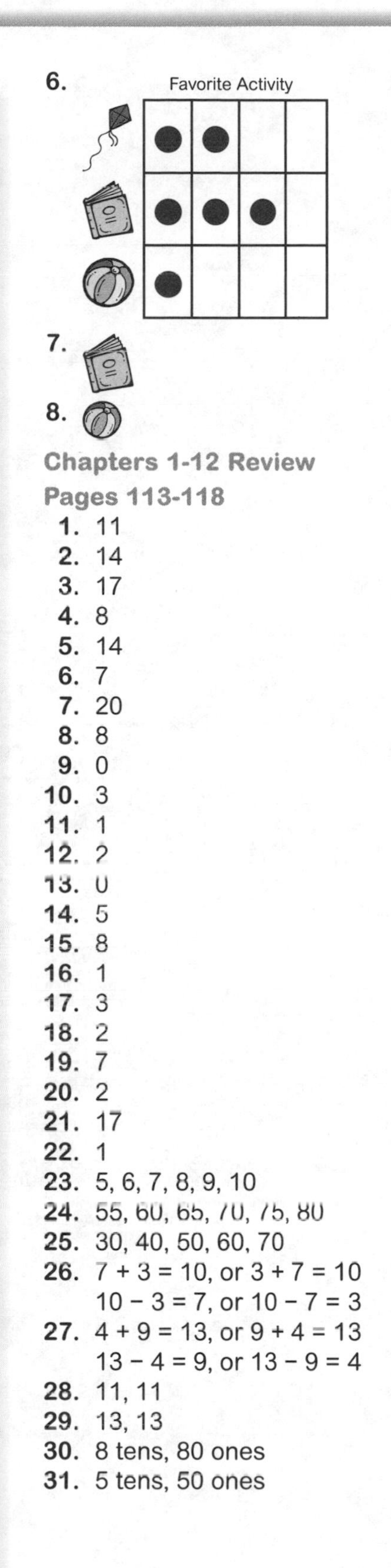

6. Favorite Activity

kite	●	●		
book	●	●	●	
ball	●			

7. (book)
8. (ball)

Chapters 1-12 Review
Pages 113-118

1. 11
2. 14
3. 17
4. 8
5. 14
6. 7
7. 20
8. 8
9. 0
10. 3
11. 1
12. 2
13. 0
14. 5
15. 8
16. 1
17. 3
18. 2
19. 7
20. 2
21. 17
22. 1
23. 5, 6, 7, 8, 9, 10
24. 55, 60, 65, 70, 75, 80
25. 30, 40, 50, 60, 70
26. 7 + 3 = 10, or 3 + 7 = 10
10 − 3 = 7, or 10 − 7 = 3
27. 4 + 9 = 13, or 9 + 4 = 13
13 − 4 = 9, or 13 − 9 = 4
28. 11, 11
29. 13, 13
30. 8 tens, 80 ones
31. 5 tens, 50 ones
32. equal to, =
33. less than, <
34. 77
35. 91
36. 20
37. 86, 66
38. 69, 49
39. 53
40. 60
41.  (monkey)
42. (car)
43. 11:00
44. 3:30
45. 7:30
46. 7:30
47. 9:00
48. 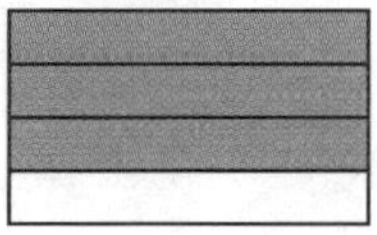 (3 of 4 parts shaded)
49. 4
50. 1